Lecture Notes in Mathematics

A collection of informal reports and seminars

Edited by A. Dold, Heidelberg and B. Eckmann, Zürich

47

J. Bénabou, R. Davis, A. Dold
J. Isbell, S. MacLane,
U. Oberst, J.-E. Roos

Reports of the Midwest Category Seminar

1967

Springer-Verlag · Berlin · Heidelberg · New York

Library of Congress Catalog Card Number 67 - 31230. Title No. 7367

Contents

INTRODUCTION TO BICATEGORIES

Jean Bénabou

Part I

<u>Introduction.</u> This is the first part of a work concerned with the study of the following type of structure: A family of categories $\underline{S}(A, B)$ $(A, B$ in a set $\underline{S}_o)$ together with pairing functors $c(A, B, C)$: $\underline{S}(A, B) \times \underline{S}(B, C) \to \underline{S}(A, C)$ which up to given coherent isomorphisms behave as if the $\underline{S}(A, B)$ were the $\mathrm{Hom}_?(B, A)$ for some "category" ?. The best known cases are perhaps $\underline{S}_o$ = one point, then we have a single category $\underline{S}$ with a multiplication in the sense of [B. 1], or a 2-category [B. 3] where the associativity isomorphisms are identities, or $\underline{S}_o$ = a set of rings, $\underline{S}(A, B)$ = category of (A, B)-Bimodules and $c(A, B, C) = \otimes_B$.

In § 1 we formalise this situation in the definition of bicategory and show in § 2 that many other cases considered by Epstein [E] or Yoneda [Y] fit in this pattern.

Even more important is the notion of morphisms defined in § 4 where we do not require the functors $F(A, B)$: $\underline{S}(A, B) \to \overline{\underline{S}}(\overline{A}, \overline{B})$ to commute with the $c(A, B, C)$, not even up to isomorphisms. The justification for such an apparently too complicated and unnecessarily general definition is in the number of examples (see § 5) ranging from monads to pseudo-functors of [Gr] which can be handled and in the fact that most of the results ex-

---

This research was supported in part by a grant from the Office of Naval Research.

pected for the strict homomorphisms, hold for general morphisms, and have meaningful interpretation (§ 6).

In § 7 we define some of the invariants of a bicategory: the Poincaré and classifying categories and the Picard groupoid which will be used in Part II. Finally § 8 is devoted to the construction of the analogue of the path space, namely the bicategory of cylinders, which gives the possibility to define transformations between morphisms (similar to natural transformations, or homotopies). For this construction we have used heavily the geometrical analogy without which definitions and results seem artificial and are incomprehensible. In many cases we have even replaced the proofs -- essentially setting up very big commutative diagrams-- by more suggestive pictures.

In Part II, we will first complete the construction of the 3-dimensional part of Bicat, by defining "modifications" between transformations, then study the notions of representability, adjointness and equivalence, which are quite different in the two-dimensional case from their ordinary analogue. Then we will examine the case when the functors $c(A, B, C)$ have a right adjoint, and finally study many examples of bicategories, devoting the greatest time to bicategories of "Profunctors".

§1. Bicategories

(1.1) Local definition. A bicategory $\underline{S}$ is determined by the following data:

(i) A set $\underline{S}_0 = Ob(\underline{S})$ called set of objects, or vertices of $\underline{S}$.

(ii) For each pair (A, B) of objects, a category $\underline{S}(A, B)$.

An object S of $\underline{S}(A, B)$ is called an edge or arrow of $\underline{S}$, and written $A \overset{S}{\longleftarrow} B$; the composition sign $\circ$ of maps in $\underline{S}(A, B)$ will usually be omitted. A map s from S to S' will be called a 2-cell and written $s\colon S \Longrightarrow S'$, or better, will be represented by :  , the composition will thus correspond to the pasting:

,

the identity maps of the categories $\underline{S}(A, B)$ will be called degenerate 2-cells. (We shall in particular use this representation with categories as vertices, functors as arrows and natural transformations as 2-cells in (v) and (vi) below.)

(iii) For each triple (A, B, C) of objects of $\underline{S}$, a composition functor:

$$c(A, B, C) : \underline{S}(A, B) \times \underline{S}(B, C) \longrightarrow \underline{S}(A, C).$$

We write $S \circ T$ and $s \circ t$ instead of $c(A, B, C)(S, T)$ and $c(A, B, C)(s, t)$ for (S, T) and (s, t) objects and maps of $\underline{S}(A, B) \times \underline{S}(B, C)$, and abbreviate $Id_S \circ t$ and $s \circ Id_T$ into $S \circ t$ and $s \circ T$. This composition corresponds to to the pasting:

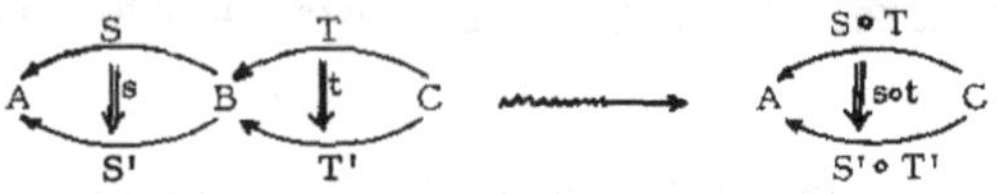

(iv) For each object A of $\underline{S}$ an object I_A of $\underline{S}(A, A)$ called identity arrow of A. The identity map of I_A in $\underline{S}(A, A)$ is denoted $i_A : I_A \Longrightarrow I_A$ and called identity 2-cell of A.

(v) For each quadruple (A, B, C, D) of objects of $\underline{S}$, a natural iso-morphism a(A, B, C, D), called associativity isomorphism, between the two composite functors bounding the diagram:

$$\underline{S}(A, B) \times \underline{S}(B, C) \times \underline{S}(C, D) \xrightarrow{\mathrm{Id} \times c(B, C, D)} \underline{S}(A, B) \times \underline{S}(B, D)$$

$$c(A, B, C) \times \mathrm{Id} \bigg\downarrow \qquad a(A, B, C, D) \qquad \bigg\downarrow c(A, B, D)$$

$$\underline{S}(A, C) \times \underline{S}(C, D) \xrightarrow{\quad c(A, C, D) \quad} \underline{S}(A, D)$$

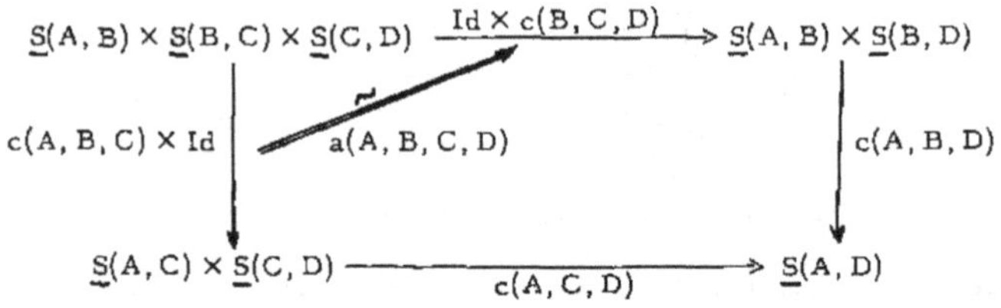

Explicitly:

$$a(A, B, C, D): c(A, C, D) \circ (c(A, B, C) \times \mathrm{Id}) \longrightarrow c(A, B, D) \circ (\mathrm{Id} \times c(B, C, D))$$

If (S, T, U) is an object of $\underline{S}(A, B) \times \underline{S}(B, C) \times \underline{S}(C, D)$ the isomorphism $a(A, B, C, D)(S, T, U): (S \bullet T) \bullet U \xrightarrow{\sim} S \bullet (T \bullet U)$ in $\underline{S}(A, D)$ is called the component of a(A, B, C, D) at (S, T, U) and is abbreviated into a(S, T, U) or even a, except when confusions are possible (cf. §3 for example).

(vi) For each pair (A, B) of objects of $\underline{S}$, two natural isomorphisms $\ell(A,B)$ and $r(A,B)$, called <u>left</u> and <u>right</u> identities, between the functors bounding the diagrams:

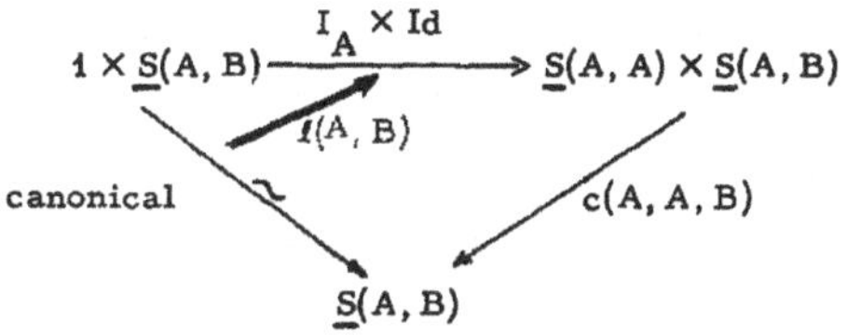

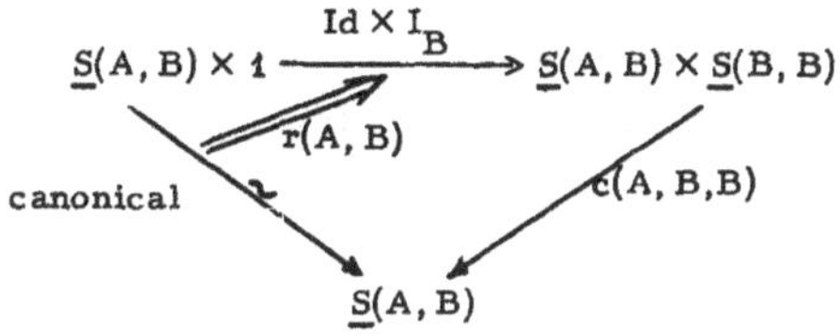

If S is an object of $\underline{S}(A,B)$, the isomorphism, component at S of $\ell(A,B)$,

$$\ell(A,B)(S):\ I_A \circ S \xrightarrow{\sim} S$$

is abbreviated into $\ell(S)$ or even ℓ, and similarly we write:

$$r = r(S) = r(A,B)(S):\ S \circ I_B \xrightarrow{\sim} S.$$

The families of natural isomorphisms $a(A,B,C,D)$, $\ell(A,B)$ and $r(A,B)$ are furthermore required to satisfy the following axioms:

(A. C.) Associativity coherence: If (S,T,U,V) is an object of $\underline{S}(A,B) \times \underline{S}(B,C) \times \underline{S}(C,D) \times \underline{S}(D,E)$ the following diagram commutes:

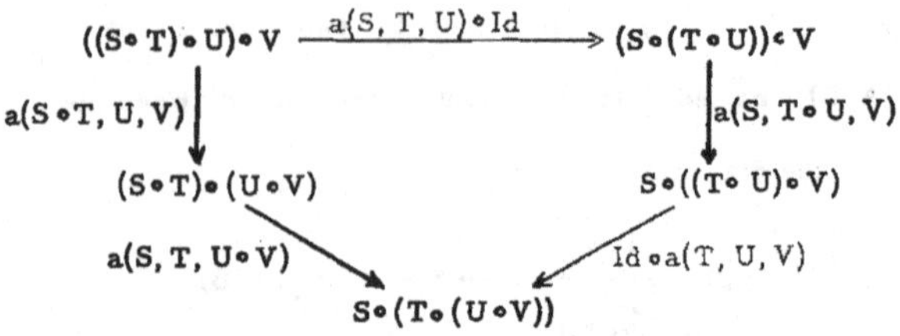
((S∘T)∘U)∘V
a(S, T, U)∘Id
(S∘(T∘U))∘V
a(S∘T, U, V)
a(S, T∘U, V)
(S∘T)∘(U∘V)
S∘((T∘U)∘V)
a(S, T, U∘V)
Id∘a(T, U, V)
S∘(T∘(U∘V))

(I. C.) Identity coherence: If (S, T) is an object of $\underline{S}(A, B) \times \underline{S}(B, C)$ the following diagram commutes:

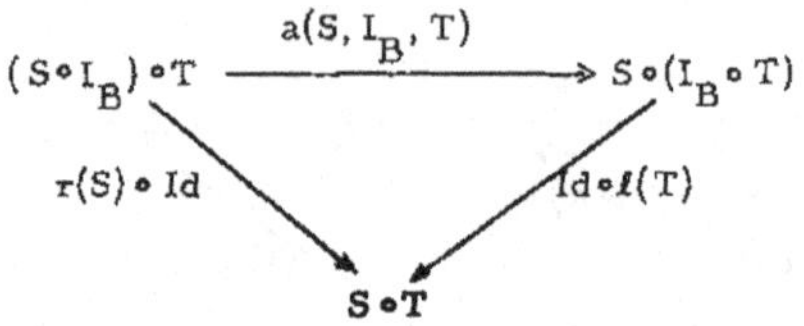
a(S, I_B, T)
(S∘I_B)∘T
S∘(I_B∘T)
r(S)∘Id
Id∘ℓ(T)
S∘T

<u>(1.2) Remark:</u> In order to avoid cumbersome notations, when the $\underline{S}(A, B)$'s shall not be disjoint, we will identify them with their canonical images in the disjoint union.

<u>(1.3) Global definition</u>

(1.3.1) A <u>bigraph</u> (or <u>bidiagram scheme</u>) Σ is a diagram of sets and maps

$$\Sigma_o \underset{\partial_1^{(1)}}{\overset{\partial_o^{(1)}}{\leftleftarrows}} \Sigma_1 \underset{\partial_1^{(2)}}{\overset{\partial_o^{(2)}}{\leftleftarrows}} \Sigma_2 \quad , \qquad \text{such that:}$$

(1.3.2) $$\partial_i^{(1)} \partial_o^{(2)} = \partial_i^{(1)} \partial_1^{(2)} \qquad (i = 0, 1) .$$

We usually omit the superscript. Elements of $\Sigma_o, \Sigma_1, \Sigma_2$ are called vertices $A, B, \ldots$, arrows $S, T, \ldots$, and 2-cells $s, t, \ldots$. A 2-cell is represented by:

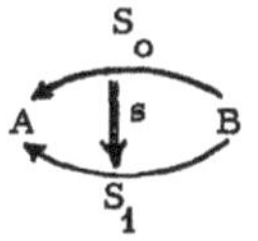

with : $S_o = \partial_o s \qquad S_1 = \partial_1 s$

$$\partial_o S_o = B = \partial_o S_1 \quad , \partial_1 S_o = A = \partial_1 S_1$$

For $n = 0, 1, 2$ we call <u>n-skeleton</u> the set $\Sigma^{[n]} = \bigcup_{i=0}^{n} \Sigma_i$. A bicategory $\underline{S}$ admits obviously an underlying bidiagram, which we usually also write $\underline{S}$, thus $S^{[k]}$ $k = 0, 1, 2$ makes sense. In particular, $Cat^{[1]}$ consists of "all" categories and functors (see (2. 2)).

(1. 3. 3) A <u>multiplication</u> μ on a bigraph Σ consists of maps:

$$\mu^{(2)} : \Sigma_2 \times_{\Sigma_1} \Sigma_2 \longrightarrow \Sigma_2 \qquad (s_1, s_2) \rightsquigarrow s_1 s_2$$

$$\mu^{(1)} : \Sigma_1 \times_{\Sigma_o} \Sigma_1 \longrightarrow \Sigma_1 \qquad (S, T) \rightsquigarrow S \circ T$$

$$\mu^{(2)} : \Sigma_2 \times_{\Sigma_o} \Sigma_2 \longrightarrow \Sigma_2 \qquad (s, t) \rightsquigarrow s \circ t$$

(by (1.3.2) there are only two maps from Σ_2 to Σ_o , thus $\Sigma_2 \times_{\Sigma_o} \Sigma_2$ is well defined) such that the diagrams (1.3.4) and (1.3.5) commute:

(1.3.4)
$$\begin{array}{ccccc} \Sigma_2 & \xleftarrow{pr_1} & \Sigma_2 \times_{\Sigma_1} \Sigma_2 & \xrightarrow{pr_2} & \Sigma_2 \\ \downarrow{\scriptstyle \partial_1} & & \downarrow{\scriptstyle \mu^{(2)}} & & \downarrow{\scriptstyle \partial_o} \\ \Sigma_1 & \xleftarrow[\partial_1]{} & \Sigma_2 & \xrightarrow[\partial_o]{} & \Sigma_1 \end{array} .$$

where $pr_i(s_1, s_2) = s_i$. That is, $\partial_o(s_o s_1) = \partial_o s_1$ and $\partial_1(s_o s_1) = \partial_1 s_o$.

And:

$$(1.3.5)\qquad \begin{array}{ccccc} \Sigma_1 \times_{\Sigma_0} \Sigma_1 & \xleftarrow{\partial_0 \times_{\Sigma_0} \partial_0} & \Sigma_2 \times_{\Sigma_0} \Sigma_2 & \xrightarrow{\partial_1 \times_{\Sigma_0} \partial_1} & \Sigma_1 \times_{\Sigma_0} \Sigma_1 \\ \downarrow{\scriptstyle \mu.^{(1)}} & & \downarrow{\scriptstyle \mu.^{(2)}} & & \downarrow{\scriptstyle \mu.^{(1)}} \\ \Sigma_1 & \xleftarrow[\partial_0]{} & \Sigma_2 & \xrightarrow[\partial_1]{} & \Sigma_1 \end{array}$$

That is: $\partial_0(s \bullet t) = (\partial_0 S \circ \partial_0 T)$; $\partial_1(s \circ t) = (\partial_1 S \circ \partial_1 T)$.

(1.3.6) A <u>degeneracy</u> (system) σ on a bigraph Σ consists of a pair of maps:

$$\Sigma_0 \xrightarrow{\sigma^{(1)}} \Sigma_1 \xrightarrow{\sigma^{(2)}} \Sigma_2$$

written $A \rightsquigarrow I_A = \sigma^{(1)}A$ and $S \rightsquigarrow i_S = \sigma^{(2)}S$, satisfying:

(1.3.7) $\qquad \partial_j^{(i)}\sigma^{(i)} = \mathrm{Id} \qquad i = 1, 2 \ ; \quad j = 0, 1 .$

Let Σ be a bigraph equipped with a multiplication μ. An <u>association</u> on (Σ, μ) is a map:

$$a : \Sigma_1 \times_{\Sigma_0} \Sigma_1 \times_{\Sigma_0} \Sigma_1 \longrightarrow \Sigma_2 \qquad (S, T, U) \rightsquigarrow a(S, T, U)$$

making commutative the diagram:

$$(1.3.8)\qquad \begin{array}{ccccc} \Sigma_1 \times_{\Sigma_0} \Sigma_1 & \xleftarrow{\mu.^{(1)} \times_{\Sigma_0} \mathrm{Id}} & \Sigma_1 \times_{\Sigma_0} \Sigma_1 \times_{\Sigma_0} \Sigma_1 & \xrightarrow{\mathrm{Id} \times_{\Sigma_0} \mu.^{(1)}} & \Sigma_1 \times_{\Sigma_0} \Sigma_1 \\ \downarrow{\scriptstyle \mu.^{(1)}} & & \downarrow{\scriptstyle a} & & \downarrow{\scriptstyle \mu.^{(1)}} \\ \Sigma_1 & \xleftarrow[\partial_0]{} & \Sigma_2 & \xrightarrow[\partial_1]{} & \Sigma_1 \end{array}$$

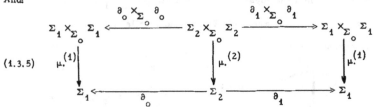

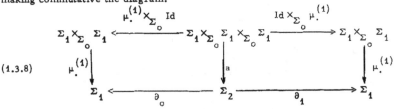

That is: $\partial_o a(S,T,U) = (S\circ T)\circ U$; $\partial_1 a(S,T,U) = S\circ(T\circ U)$.

If furthermore Σ is equipped with a degeneracy σ, <u>left</u> and <u>right</u> <u>identity (systems)</u> are maps:

$$\ell : \Sigma_1 \longrightarrow \Sigma_2 \qquad S \rightsquigarrow \ell(S)$$

$$r: \Sigma_1 \longrightarrow \Sigma_2 \qquad S \rightsquigarrow r(S)$$

making commutative the diagram:

(1.3.9)
$$\begin{array}{ccccccccc}
 & & \Sigma_1 & \xrightarrow{(\sigma^{(1)},\, Id)} & \Sigma_1 \times_{\Sigma_o} \Sigma_1 & \xrightarrow{(Id,\, \sigma^{(1)})} & \Sigma_1 & & \\
 & \swarrow^{Id}_{\sim} & \downarrow r & & \downarrow \mu^{(1)} & & \downarrow \ell & \searrow^{Id}_{\sim} & \\
\Sigma_1 & \xrightarrow[\partial_1]{} & \Sigma_2 & \xrightarrow[\partial_o]{} & \Sigma_1 & \xrightarrow[\partial_o]{} & \Sigma_2 & \xrightarrow[\partial_1]{} & \Sigma_1
\end{array}$$

That is, $\partial_o \ell(S) = I_A \circ S$; $\partial_1 \ell(S) = S$; $\partial_o r(S) = S\circ I_B$; $\partial_1 r(S) = S$.

The choice of notation is such that, if $\underline{S}$ is a bicategory, the underlying bigraph is clearly equipped with a canonical multiplication, degeneracy, association and identities, called underlying to $\underline{S}$.

In terms of $(\Sigma, \mu, \sigma, a, \ell, r)$ bicategories can be characterized by means of the following:

<u>(1.3.10) Proposition:</u> Let Σ be a bigraph equipped with μ, σ, a, ℓ, r. These data are underlying to a bicategory, then necessarily unique, iff they satisfy conditions (i) to (x) below:

(i) The following diagram is commutative:

$$\begin{array}{ccc}
\Sigma_2 \times_{\Sigma_1} \Sigma_2 \times_{\Sigma_1} \Sigma_2 & \xrightarrow{\mu^{(2)} \times_{\Sigma_1} Id} & \Sigma_2 \times_{\Sigma_1} \Sigma_2 \\
{\scriptstyle Id \times_{\Sigma_1} \mu^{(2)}} \downarrow & & \downarrow {\scriptstyle \mu^{(2)}} \\
\Sigma_2 \times_{\Sigma_1} \Sigma_2 & \xrightarrow[\mu^{(2)}]{} & \Sigma_2
\end{array} .$$

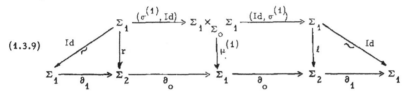

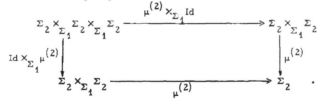

(ii) The following diagram is commutative:

$$\begin{array}{ccccc}
\Sigma_2 & \xrightarrow{(\sigma\partial_1,\, \mathrm{Id})} & \Sigma_2 \times_{\Sigma_1} \Sigma_2 & \xleftarrow{(\mathrm{Id},\, \sigma\partial_0)} & \Sigma_2 \\
& {\scriptstyle \mathrm{Id}}\searrow^{\sim} & \downarrow {\scriptstyle \mu^{(2)}} & {}^{\sim}\swarrow{\scriptstyle \mathrm{Id}} & \\
& & \Sigma_2 & &
\end{array}$$

(iii) The following diagram is commutative:

$$\begin{array}{ccccc}
(\Sigma_2 \times_{\Sigma_0} \Sigma_2) \underset{\Sigma_1 \times_{\Sigma_0} \Sigma_1}{\times} (\Sigma_2 \times_{\Sigma_0} \Sigma_2) & & \xrightarrow[\sim]{\tau} & & (\Sigma_2 \times_{\Sigma_1} \Sigma_2) \times_{\Sigma_0} (\Sigma_2 \times_{\Sigma_1} \Sigma_2) \\
\downarrow {\scriptstyle \mu.^{(2)} \times_{\Sigma_1} \mu.^{(2)}} & & & & \downarrow {\scriptstyle \mu^{(2)} \times_{\Sigma_0} \mu^{(2)}} \\
\Sigma_2 \times_{\Sigma_1} \Sigma_2 & \xrightarrow[\mu^{(2)}]{} & \Sigma_2 & \xleftarrow[\mu.^{(2)}]{} & \Sigma_2 \times_{\Sigma_0} \Sigma_2
\end{array} ,$$

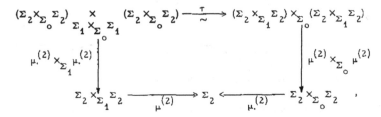

where τ is the canonical map $((s_1, t_1), (s_0, t_0)) \rightsquigarrow ((s_1, s_0), (t_1, t_0))$ defined when s_i, t_i satisfy the incidence relations depicted by:

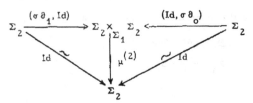

and $\mu.^{(2)} \times_{\Sigma_1} \mu.^{(2)}$ is the unique map, which exists because of (1.3.5), making commutative the diagram:

$$\begin{array}{ccc}
(\Sigma_2 \times_{\Sigma_0} \Sigma_2) \underset{\Sigma_1 \times_{\Sigma_0} \Sigma_1}{\times} (\Sigma_2 \times_{\Sigma_2} \Sigma_2) & \hookrightarrow & (\Sigma_2 \times_{\Sigma_0} \Sigma_2) \times (\Sigma_2 \times_{\Sigma_0} \Sigma_2) \\
\downarrow {\scriptstyle \mu.^{(2)} \times_{\Sigma_1} \mu.^{(2)}} & & \downarrow {\scriptstyle \mu.^{(2)} \times \mu.^{(2)}} \\
\Sigma_2 \times_{\Sigma_1} \Sigma_2 & \hookrightarrow & \Sigma_2 \times \Sigma_2
\end{array} ,$$

where the horizontal arrows are the canonical monomorphisms of pull-backs into products.

(iv) The following diagram is commutative:

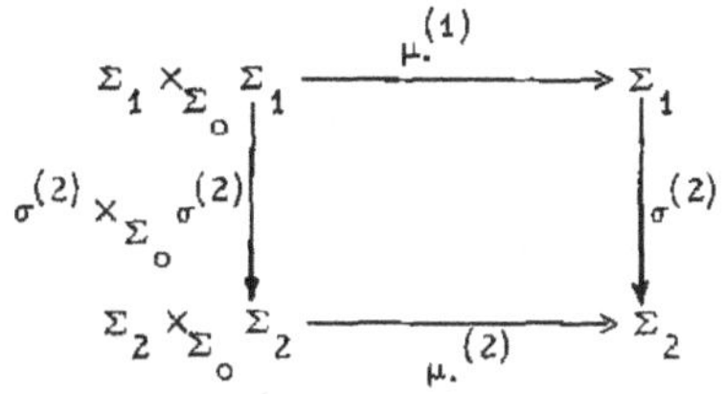

From (1.3.8) and (1.3.5) it follows that the exterior of the following diagram is commutative, thus there exists a unique map φ making the whole diagram commutative

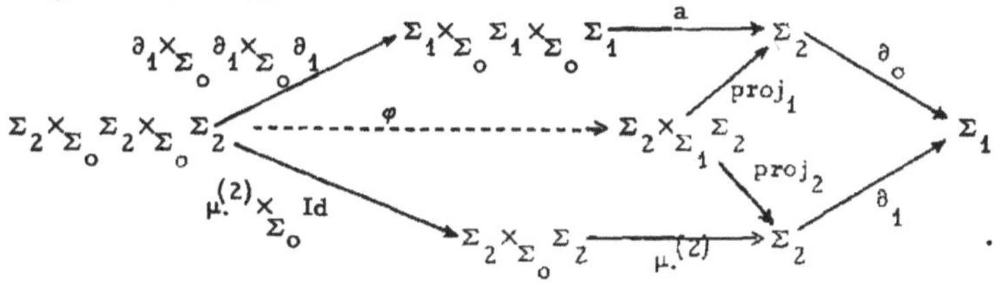

Similarly, let φ' be the unique map making commutative:

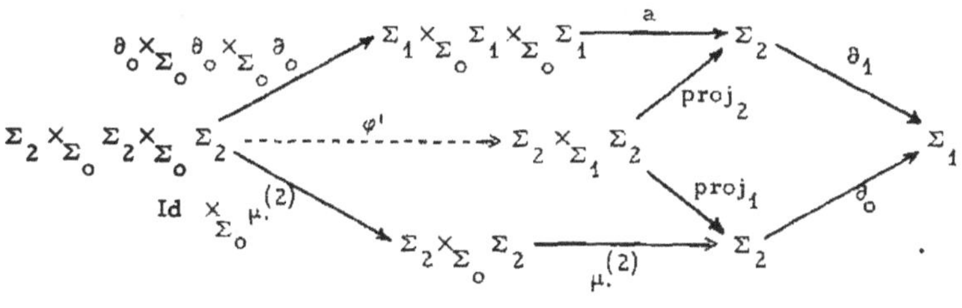

(v) The following diagram is commutative:

$$\begin{array}{ccc}
\Sigma_2 \times_{\Sigma_0} \Sigma_2 \times_{\Sigma_0} \Sigma_2 & \xrightarrow{\varphi} & \Sigma_2 \times_{\Sigma_1} \Sigma_2 \\
\downarrow{\scriptstyle \varphi'} & & \downarrow{\scriptstyle \mu^{(2)}} \\
\Sigma_2 \times_{\Sigma_1} \Sigma_2 & \xrightarrow[\mu^{(2)}]{} & \Sigma_2
\end{array}$$

From (1.3.9) it follows that the exterior of the following diagram commutes, hence there is a unique ψ_ℓ making the whole diagram commute:

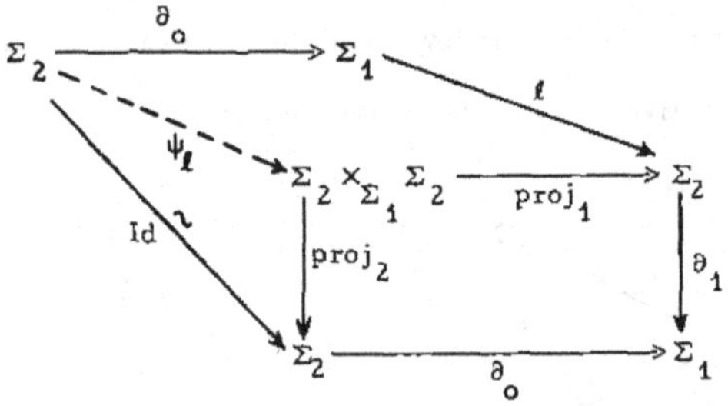

Similarly let ψ'_ℓ be the unique map making commutative:

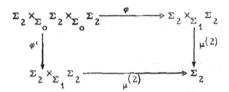

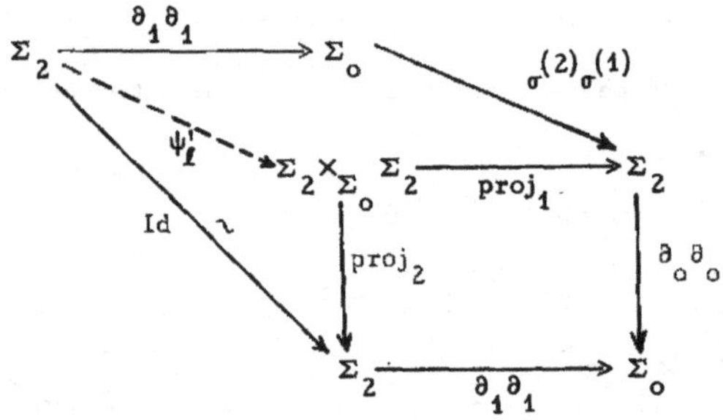

And ψ_ℓ'' the unique map making commutative:

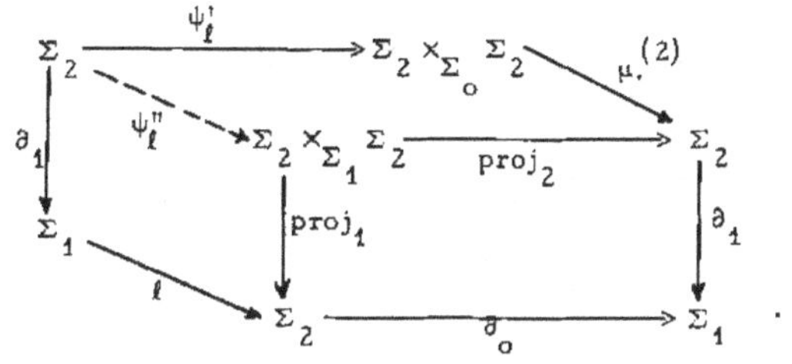

$(vi)_\ell$: The following diagram is commutative:

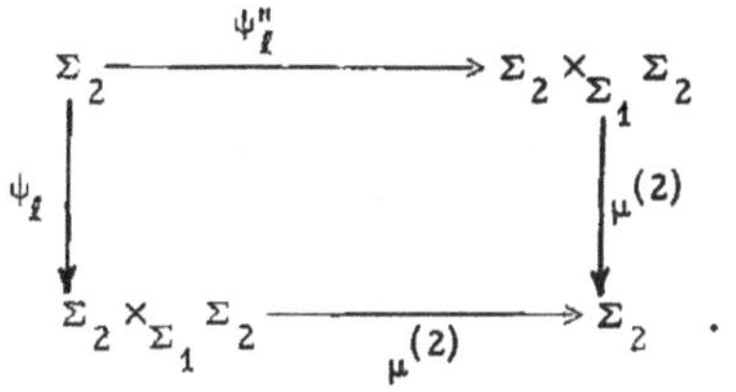

We let the reader write the analogous diagram $(vi)_r$ for r.

From the definitions of $a, \mu.^{(2)}$ and $\mu.^{(1)}$ it follows that the exterior of the diagram below is commutative, hence there is a unique map θ_1 making the whole diagram commutative:

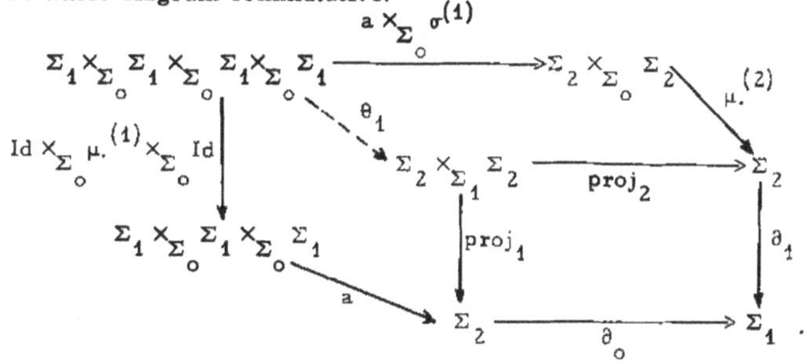

In a similar way, let θ_2 be the unique map making commutative the diagram:

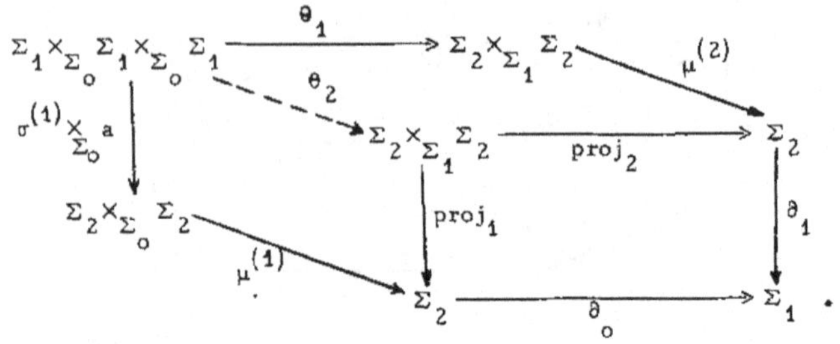

and θ_3 be the unique map making commutative:

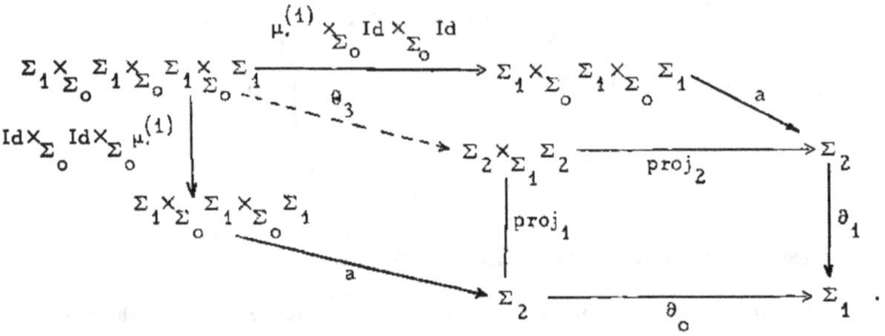

(vii) The following diagram is commutative:

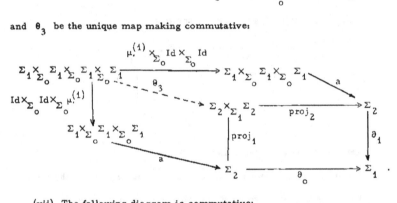

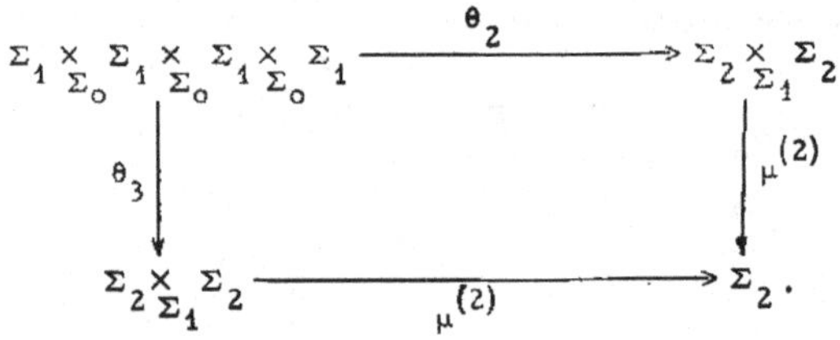

Again one can check the commutativity of the exterior of the diagram below, and define α to be the unique map making commutative:

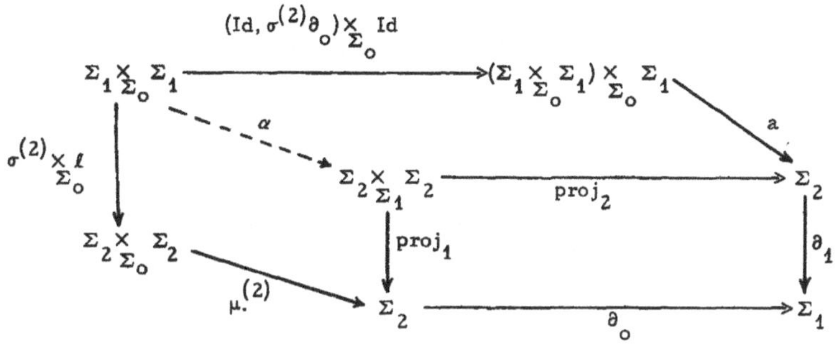

(viii) The following diagram is commutative:

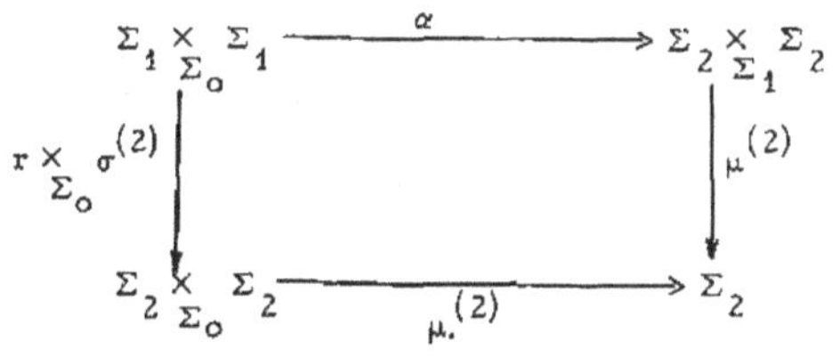

(ix) There exists a map, (necessarily unique)

$$\bar{a}\colon \Sigma_1 \underset{\Sigma_o}{\times} \Sigma_1 \underset{\Sigma_o}{\times} \Sigma_1 \longrightarrow \Sigma_2$$

such that $\partial_o \bar{a} = \partial_1 a$, $\partial_1 \bar{a} = \partial_o a$, making commutative the diagram (where a map into a pullback is denoted by its two components):

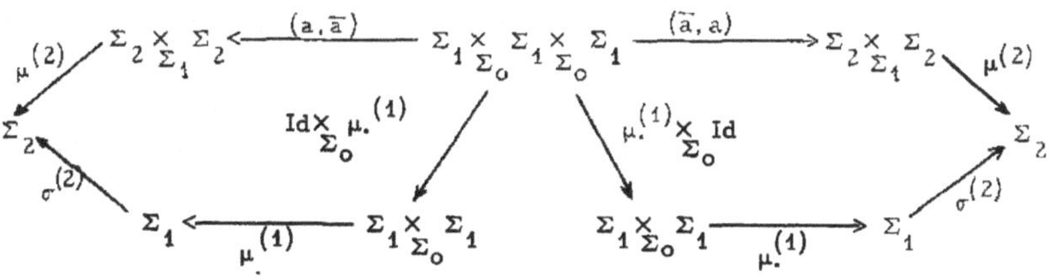

$(x)_\ell$ There exists a map, again unique

$$\bar{\ell}: \Sigma_1 \longrightarrow \Sigma_2$$

such that $\partial_0 \bar{\ell} = \partial_1 \ell$, $\partial_1 \bar{\ell} = \partial_0 \ell$, making commutative:

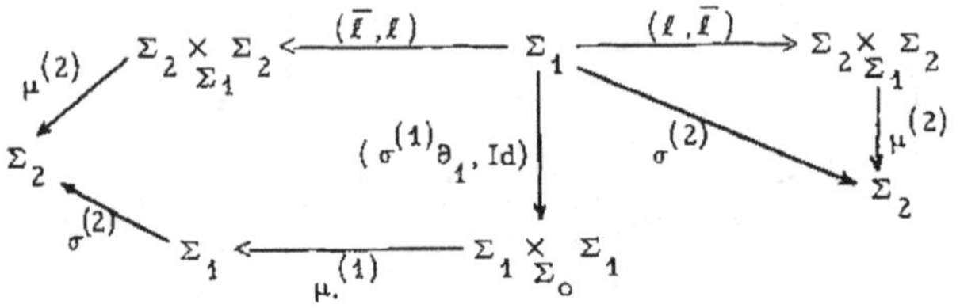

Again there is a similar diagram $(x)_r$ for r, left to the reader to provide.

We make the following comments about the proof, the details of which are left to the reader. Given $(\Sigma, \mu, \dots)$ satisfying (i) to (x), we define a bicategory $\underline{S}$ as follows:

It has Σ_0 as set of objects. If $A, B \in \Sigma_0$, $\underline{S}(A, B)$ has as objects the elements S of Σ_1 such that $\partial_1 S = A$, $\partial_0 S = B$ and as arrows the elements s of Σ_2 such that $\partial_0\partial_0 s = B$, $\partial_1\partial_1 s = A$; the domain and codomain of s are $\partial_0 s$ and $\partial_1 s$, the composition is $(s_1, s_2) \rightsquigarrow s_1 s_2$; the conditions (i) and (ii) state precisely that $\underline{S}(A, B)$ is a category. For each triple (A, B, C), c(A, B, C) is given on objects (resp. maps) by the restrictions of $\mu.^{(1)}$ (resp. $\mu.^{(2)}$) to $\mathrm{Ob}\,\underline{S}(A, B) \times \mathrm{Ob}\,\underline{S}(B, C) \subset \Sigma_1 \times_{\Sigma_0} \Sigma_1$ (resp. ...). Then (iii) and (iv) mean that the c(A, B, C)'s are bifunctors. For each

A, B, C, D, the restriction of a to $\mathrm{Ob}\,\underline{S}(A,B) \times \mathrm{Ob}\,\underline{S}(B,C) \times \mathrm{Ob}\,\underline{S}(C,D) \subset \Sigma_1 \underset{\Sigma_0}{\times} \Sigma_1 \underset{\Sigma_0}{\times} \Sigma_1$ is a(A, B, C, D) whose naturality follows from (v). For each A, B the restriction of ℓ to $\mathrm{Ob}\,\underline{S}(A,B) \subset \Sigma_1$ is according to $(vi)_\ell$ a natural transformation $\ell(A,B)$ (same thing for r). The coherence (A.C) is expressed by (vii), (I.C) by (vii), and finally (ix) and $(x)_\ell$ state that the a's and ℓ's are isomorphisms with inverses the restrictions of $\bar{a}$ and $\bar{\ell}$ (and $(x)_r$ gives r^{-1}).

(1.4) Remark: The proposition (1.3.10) makes it possible to define a bicategory in terms of $\Sigma_0, \Sigma_1, \Sigma_2$, the maps $\partial, \mu, a, \ldots$ satisfying (i) to (x). This definition, although much longer and less intuitive than (1.1) has the following advantages:

(i) It is purely "diagrammatic", and can be stated with Σ_i objects of any category with pullbacks, giving such examples as topological, or ordered, bicategories. (The only place where elements were used was, for the sake of brevity, in defining τ which obviously exists in any category with pullbacks.)

(ii) Even in the case of sets, it shows that bicategories are "algebraic", i. e., defined in terms of finite inverse limits, and makes available all the general theorems on algebraic structures (see e. g. (7.4.1) below).

2. Examples of bicategories

The following examples are designed to fix the terminology for further reference and, hopefully, to provide the reader with intuitive support and motivation for the forthcoming nonsense:

(2.1) 2-Categories: A 2-category is defined in [B. 3] (example 2) by data identical to (i), (ii), (iii), and (iv) of (1.1), such that the diagrams of functors bounding the 2-cells of (v) and (vi) are commutative. If we take $a(A,B,C,D)$, $\ell(A,B)$, $r(A,B)$ to be the identity natural isomorphisms, the axioms (A. C) and (I. C) are obviously satisfied. Thus, the 2-categories (also called Hypercategories in [E. K]) can be identified with the bicategories where c is strictly associative, with I_A's as strict identities for c; and a, ℓ, r are identity natural transformations. We shall see however that the notion of morphisms of bicategories, even when restricted to 2-categories, gives a wider and more interesting class than the 2-functors (cf. (5.3)).

In particular, we will denote by Tac(*) the 2-category with objects "all" categories, Tac(A, B) being the category of all functors from B to A, and if S, S', S", T, T' and s, s', t are functors and natural transformations satisfying the incidence relations represented in the "bidiagram"

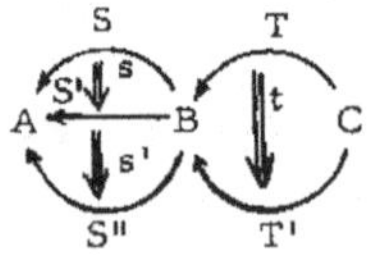

,

(*) The notation Moh(A, B) for Hom(B, A), A, B objects of any category, is due to Epstein. See (3.4.1) for the "transpose" Cat of Tac.

then $S \circ T$ is the composite functor ST, $s' \circ s$ is the usual composite of natural transformations, and $s \circ t = (S' * t) \circ (s * T) = (s * T') \circ (S * t)$ with the notatation of [G] (p. 269). In this case we shall write $s * t$ instead of $s \circ t$.

<u>(2. 2) Multiplicative categories.</u> Let $\underline{M} = (\underline{A}, \otimes, \Lambda, \theta, \gamma, \delta)$ be a category with multiplication (c. m.) as defined in [B. 1]. Take $\underline{S}_0$ to be a set having a single element, say 0. Define $\underline{S}(0, 0) = \underline{A}$, $c(0,0,0) = \otimes$, $I_0 = \Lambda$, $a(0,0,0,0) = \theta$, $\ell(0,0) = \gamma$, $r(0,0) = \delta$. These data satisfy (A. C) and (I. C) and thus define a bicategory $\underline{S}$ with one object. Conversely, every bicategory with one object "is" a c. m. . More generally we have:

<u>Proposition (2. 2. 1).</u> Let $\underline{S}$ be a bicategory and A an object of $\underline{S}$, then $c(A,A,A) = \otimes$, $I_A = \Lambda$, $a(A,A,A,A) = \theta$, $\ell(A,A) = \gamma$, $r(A,A) = \delta$ determine on the category $\underline{A} = \underline{S}(A,A)$ a multiplicative structure called induced by $\underline{S}$.

The proposition follows from the general coherence theorem of [B. 4]. See also [M], and compare with the "one-dimensional" case: a monoid "is" a category with one object, and for any category $\underline{C}$ and object A of $\underline{C}$, Hom(A, A) is a monoid.

Note furthermore that if $\underline{S}$ is a 2-category, then $\underline{S}(A,A)$ is a strictly associative c. m. . In particular, taking $\underline{S}$ = Tac, we get the multiplicative structure of the category of endofunctors of any category defined in [B. 1].

(2.3) Actions of Multiplicative Categories. Let $\underline{M}$ be a c.m. and $\underline{X}$ any category. A left action of $\underline{M}$ on $\underline{X}$ is defined by:

(i) A functor: $\underline{A} \times \underline{X} \longrightarrow \underline{X}$; $(A, X) \rightsquigarrow A \otimes X$,

(ii) Natural isomorphisms:

$$\alpha : (A_1 \otimes A_2) \otimes X \xrightarrow{\sim} A_1 \otimes (A_2 \otimes X) \quad \text{and} \quad \eta : \Lambda \otimes X \xrightarrow{\sim} X,$$

satisfying "obvious" coherence conditions.

Such a left action can be identified with the bicategory $\underline{S}$ described by: $\mathrm{Ob}(\underline{S}) = \{0, 1\}$, $\underline{S}(0,0) = \underline{A}$, $\underline{S}(0,1) = \underline{X}$, $\underline{S}(1,1) = 1$, $\underline{S}(0,1) = \mathbf{0}$, $c(0,0,0) = \otimes$, $c(0,0,1) = \otimes$, ... The reader will provide the rest of the data.

Conversely, if $\underline{S}$ is any bicategory, and A, B are two objects of $\underline{S}$, the c.m. $\underline{S}(A, A)$ inherits from $\underline{S}$ a canonical left action on $\underline{S}(A, B)$ given by: $S \circ T = S \otimes T$, $\alpha = a : (S_1 \otimes S_2) \otimes T \xrightarrow{\sim} S_1 \otimes (S_2 \otimes T)$, $\eta = \ell$, $I_A \otimes T \xrightarrow{\sim} T$.

For example, if $\underline{X}$ is any category and $\underline{M} = \mathrm{Tac}(\underline{X}, \underline{X})$ is the category of endofunctors of $\underline{X}$, it acts on $\underline{X}$ by $(F, X) \rightsquigarrow F(X)$. Or again: If $\underline{X}$ is any abelian category with arbitrary colimits (resp. any category with arbitrary products) and $\underline{M}$ is the c.m. of abelian groups (resp. Sets*) with $\otimes$ (resp $\times$) as multiplication, a choice of colimits (resp. products) determines a canonical left action by $(A, X) \rightsquigarrow A \otimes X$ (resp. $(A, X) \rightsquigarrow X^A = \prod_{a \in A} X$).

Similarly, we can define a right action, or a biaction: $\underline{M}$ and $\overline{\underline{M}}$ are c. m., $\underline{M}$ acts on the left on $\underline{X}$, $\overline{\underline{M}}$ on the right, and both actions "commute" up to coherent isomorphisms:

$$(A\otimes X)\otimes\overline{A} \xrightarrow{\sim} A\otimes(X\otimes\overline{A}).$$

All these data and axioms can be reduced to: a bicategory $\underline{S}$ with two objects, say 0 and 1, such that $\underline{S}(1,0) = \mathbb{0}$.

(2.4). In [E], Epstein considers the following situation: Categories $\underline{A}, \underline{B}, \underline{C}, \underline{M}, \underline{N}, \underline{O}$, functors, denoted by $\otimes$:

$$\underline{A}\times\underline{B} \longrightarrow \underline{M}\ ;\ \underline{B}\times\underline{C}\longrightarrow\underline{N}\ ;\ \underline{M}\times\underline{C}\longrightarrow\underline{O}\ ;\ \underline{A}\times\underline{N}\longrightarrow\underline{O}$$

and a natural isomorphism $\alpha:(A\otimes B)\otimes C \xrightarrow{\sim} A\otimes(B\otimes C)$. This reduces to: A bicategory $\underline{S}$ having four objects 0, 1, 2, 3, such that

$$\underline{S}(i,j) = \mathbb{0} \quad \text{for } i > j \quad \text{and} \quad \underline{S}(i,i) = 1.$$

(Take then $\underline{S}(0,1) = \underline{A}$, $\underline{S}(1,2) = \underline{B}$, $\underline{S}(2,3) = \underline{C}$, $\underline{S}(0,2) = \underline{M}$, $\underline{S}(1,3) = \underline{N}$, $\underline{S}(0,3) = \underline{O}$, etc.)

(2.5) Bimodules. A basic example, to be generalized and studied in Part II, is the bicategory Bim of bimodules defined as follows: The objects are the rings with identity. If A and B are rings, $\mathrm{Bim}(A,B) = {}_A\underline{M}_B$ is the category of (A, B)-bimodules. If $M \in {}_A\underline{M}_B$ and $N \in {}_B\underline{M}_C$, $M\circ N$ is the (A, C)-bimodule $M\otimes_B N$. The ring A, as an (A, A)-bimodule is I_A. Finally, a, ℓ, r are the usual isomorphisms of the tensor.

With this definition, an "arrow" between two rings A and B is an (A, B)-bimodule, composition being the tensor. Note that the usual arrows, i.e., ring homomorphisms $f: B \longrightarrow A$ determine (A, B)-bimodules $M_f = A$ (viewed as (A, B)-bimodule through f) and that, if $f': B \longrightarrow A$, M_f and $M_{f'}$ are isomorphic in ${}_A\underline{M}_B$ iff $f = f'$. Thus ring homomorphisms "are" arrows of Bim. Furthermore, if $g: C \longrightarrow B$, we have obviously a canonical isomorphism $M_f \bullet M_g \longrightarrow M_{fg}$.

<u>(2.6) Spans.</u> Let $\underline{C}$ be any category with pullbacks (*). Choosing for each diagram $U \longrightarrow V \longleftarrow W$ in $\underline{C}$ a pull back diagram:

$$\begin{array}{ccc} U \prod_V W & \longrightarrow & W \\ \downarrow & & \downarrow \\ U & \longrightarrow & V \end{array}$$

we now define "the" bicategory Sp $\underline{C}$ of spans of $\underline{C}$ (another choice of pullbacks would give a bicategory isomorphic in an obvious sense). The objects of Sp$\underline{C}$ are the objects of $\underline{C}$. If A and B are two objects, the category Sp $\underline{C}$(A, B) has as objects, i.e., arrows of Sp$\underline{C}$, all diagrams $s: A \xleftarrow{\alpha} X \xrightarrow{\beta} B$ in $\underline{C}$. A map s in Sp$\underline{C}$(A, B) from S to S': $A \xleftarrow{\alpha'} X' \xrightarrow{\beta'} B$ is a commutative diagram in $\underline{C}$

$$s: \begin{array}{ccccc} & & X & & \\ & \overset{\alpha}{\swarrow} & \downarrow{\scriptstyle \xi} & \overset{\beta}{\searrow} & \\ A & & & & B \\ & \underset{\alpha'}{\nwarrow} & X' & \underset{\beta'}{\nearrow} & \end{array} .$$

(*) The notion of span was introduced by Yoneda in [Y] for the case $\underline{C} = Cat^{[1]}$, the category of categories.

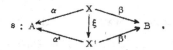

Composition in $Sp\,\underline{C}(A,B)$ is the obvious one. The pairings:

$$Sp\,\underline{C}(A,B) \times Sp\,\underline{C}(B,C) \longrightarrow Sp\,\underline{C}(A,C)\ ;\quad (S,T) \rightsquigarrow S \circ T$$

are defined by pullback. Explicitely, if $T\colon B \xleftarrow{\gamma} Y \xrightarrow{\delta} C$,

$$S \circ T \text{ is } \quad A \xleftarrow{\alpha p_1} X \times_B Y \xrightarrow{\delta p_2} C\ ,$$

where $p_1\colon X \times_B Y \longrightarrow X$ and $p_2\colon X \times_B Y \longrightarrow Y$ are the canonical projections of the pullback. The identity arrow of A is:

$I_A\colon A \xleftarrow{Id} A \xrightarrow{Id} A$. Finally, a, ℓ, r are given by the usual isomorphisms of associativity and identity of pullbacks.

Note that if $\underline{C}$ has a final object, say 1, and thus finite products, the multiplicative structure on $\underline{C}$ defined by the product, is isomorphic to $Sp\,\underline{C}(1,1)$ with the induced structure. Dually, if $\underline{C}$ has pushouts, define the bicategory $Cosp\,\underline{C}$ of Cospans in $\underline{C}$, isomorphic with $Sp(\underline{C}^*)$.

(2.7) Local properties of bicategories. Let P be a property of categories, a bicategory $\underline{S}$ is locally P if all the categories $\underline{S}(A,B)$ satisfy P. For example, Bim is locally abelian.

If in the data of $\underline{S}$ all the $\underline{S}(A,B)$ are partially ordered sets, identified to categories, the coherence conditions are automatically satisfied, thus $\underline{S}$ is a bicategory, called locally ordered. The extreme types of partially ordered sets are the discrete ($x \leq y$ iff $x = y$) and the anti-discrete (for all x and for all y, $x \leq y$).

The locally discrete bicategories, are always 2-categories. Moreover, all their 2-cells are degenerate, we will therefore call them 1-dimensional, and identify categories $\underline{C}$ with 1-dimensional bicategories (by identifying the sets $\mathrm{Hom}_{\underline{C}}(A, B)$ with discrete categories). Thus we will speak of morphisms of a category $\underline{C}$ into a bicategory $\underline{S}$ or of the empty bicategory $\mathbb{0}$, and the punctual bicategory $\mathbb{1}$ with one object $\mathbb{0}$, and $\mathbb{1}(\mathbb{0}, \mathbb{0}) = \{0\} = \mathbb{1}$, etc .

The locally antidiscrete $\underline{S}$'s can be identified with families of sets $\underline{S}(A, B)$ equipped with maps $\underline{S}(A, B) \times \underline{S}(B, C) \longrightarrow \underline{S}(A, C)$ and base points $I_A \in \underline{S}(A, A)$ satisfying no axioms, and thus are not in general 2-categories.

The bicategory $\underline{R}$ of relations provides a good example of locally ordered bicategory: The objects are sets. If A and B are two sets $\underline{R}(A, B)$ is the power set of $A \times B$, ordered by inclusion. The pairings c are given by the composition of relations and the identity of A is the image Δ_A of A under the diagonal map $\Delta: A \longrightarrow A \times A$. It is clearly a 2-category.

This example can obviously be extended by replacing the category of sets by an abelian category $\underline{A}$ to get the category of additive relations of $\underline{A}$, or by a category with finite limits and some exactness properties which we won't list.

(2.8) Extensions. Let $\underline{A}$ be an abelian category. For each integer n, let $\underline{\mathrm{Ext}}^n_{\underline{A}}(A, B)$ be the category with objects the n-fold extensions

$$S: 0 \leftarrow A \leftarrow E_1 \leftarrow \dots E_n \leftarrow B \leftarrow 0$$

and maps the ordinary maps with endpoints fixed, i. e., the commutative diagrams:

$$\begin{array}{ccccccccccccc} S & : & 0 & \longleftarrow & A & \longleftarrow & E_1 & \longleftarrow \cdots & E_n & \longleftarrow & B & \longleftarrow & 0 \\ {\scriptstyle s}\downarrow & & & & \| & & \downarrow & & \downarrow & & \| & & \\ S' & : & 0 & \longleftarrow & A & \longleftarrow & E'_1 & \longleftarrow \cdots & E'_n & \longleftarrow & B & \longleftarrow & 0 \end{array}$$

Define: (i) $\mathbf{Ext}_{\underline{A}}(A, B)$ to be the union of the categories $\mathbf{Ext}^n_{\underline{A}}(A, B)$;

(ii) the composition pairings:

$$\mathbf{Ext}_{\underline{A}}(A, B) \times \mathbf{Ext}_{\underline{A}}(B, C) \longrightarrow \mathbf{Ext}_{\underline{A}}(A, C)$$

to be the Yoneda composition of exact sequences.

(iii) For each A in $\underline{A}$, I_A to be $0 \longleftarrow A \xleftarrow{Id} A \longleftarrow 0$

(iv) The a, ℓ, r to be the identity natural isomorphisms.

We obtain thus the <u>bicategory of extensions of</u> $\underline{A}$, written $\mathbf{Ext}_{\underline{A}}$.
The same construction can be performed when $\underline{A}$ is a relative abelian category.

3. Dualities.

For a category $\underline{C}$ there is only one kind of symmetric, namely the dual $\underline{C}^*$ of $\underline{C}$. For a bicategory $\underline{S}$, there are three such, all having the same objects, arrows and 2-cells as $\underline{S}$, described as follows:

(3.1). The conjugate $\underline{S}^c$ defined by:

$$\underline{S}^c(A,B) = [\underline{S}(A,B)]^* \;;\; I_A^c = I_A$$

$$c^c(A,B,C) = [c(A,B,C)]^* : \underline{S}^c(A,B) \times \underline{S}^c(B,C) \longrightarrow \underline{S}^c(A,C)$$

$$a^c(A,B,C,D) = [a(A,B,C,D)]^{-1} \;;\; \ell^c(A,B) = [\ell(A,B)]^{-1},$$

$$r^c(A,B) = [r(A,B)]^{-1}$$

(3.2). The transpose $\underline{S}^t$, defined by:

$$\underline{S}^t(A,B) = \underline{S}(B,A) \;,\; I_A^t = I_A .$$

$c^t(A,B,C)$ makes the following diagram of functors commutative:

$$\begin{array}{ccc}
\underline{S}^t(A,B) \times \underline{S}^t(B,C) & \xrightarrow{\;c^t(A,B,C)\;} & \underline{S}^t(A,C) = \underline{S}(C,A) \\
\Big\| & & \Big\uparrow c(C,B,A) \\
\underline{S}(B,A) \times \underline{S}(C,B) & \xrightarrow[\text{canonical}]{\;\sim\;} & \underline{S}(C,B) \times \underline{S}(B,A)
\end{array}$$

$$a^t(A,B,C,D)(S,T,U) = [a(D,C,B,A)(U,T,S)]^{-1}$$

$$\ell^t(A,B)(S) = r(B,A)(S) \quad \text{and} \quad r^t(A,B)(S) = \ell(B,A)(S).$$

(3.3). The symmetric $\underline{S}^s$, defined by $\underline{S}^s = \underline{S}^{ct}$.

If we make the convention to represent objects A, arrows S and 2-cells s by A^t, S^t, s^t (resp. $A^c, \ldots$) when considered as belonging to $\underline{S}^t$ (resp. $\underline{S}^c, \ldots$) the unpalatable formulae defining the different dualities have simple geometric pictures: The typical 2-cell of $\underline{S}$: 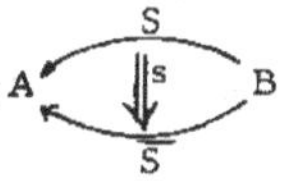

is represented in $\underline{S}^c$, $\underline{S}^t$ and $\underline{S}^s$ respectively by

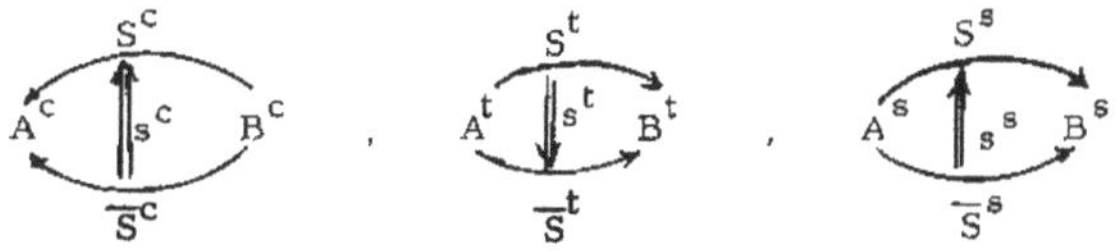

The definitions of $\circ^c, c^c, I^c, a^c, \ell^c, r^c$ (resp. ...) are "forced" by these pictures. And the equations $\underline{S}^{cc} = \underline{S}^{tt} = \underline{S}^{ss} = \underline{S}$, and $\underline{S}^{ct} = \underline{S}^{tc}$ which can be directly checked, become "geometrically obvious."

3.4 Examples.

(3.4.1). If $\underline{S}$ is a 2-category, so are $\underline{S}^c, \underline{S}^t$ and $\underline{S}^s$. In particular, the transpose Tac^t of Tac as defined in (2.1) will be denoted Cat.

(3.4.2). If $\underline{M} = (\underline{A}, \otimes, \ldots)$ is a c.m., then $\underline{M}^c$ is the category $\underline{A}^*$ dual of $\underline{A}$, equipped with "the same" multiplication as $\underline{A}$, $\underline{M}^t$ is the category $\underline{A}$ equipped with the opposite multiplication $(A, B) \rightsquigarrow B \otimes A$, and $\underline{M}^s$ is $\underline{A}^*$ with the opposite multiplication. All this obviously extends to actions of c.m. on categories (e.g., transposition transforms right action into left action ...).

(3.4.3). If $\underline{S}$ is locally ordered (discrete, antidiscrete) so are $\underline{S}^c, \underline{S}^t$, and $\underline{S}^s$. In the discrete case, i.e., when S is a category, we have furthermore: $\underline{S}^c = \underline{S}$, $\underline{S}^t = \underline{S}^s = \underline{S}^*$.

(3.4.4). Clearly every statement about bicategories contains really four statements: If a proposition P is true for $\underline{S}$, then there are conjugate, transpose and symmetric propositions P^c, P^t, P^s, true for $\underline{S}^c, \underline{S}^t$, $\underline{S}^s$ which we will omit most of the time.

4. Morphisms of bicategories

(4.1). Definition: Let $\underline{S} = (\underline{S}_0, c, I, a, l, r)$ and $\overline{\underline{S}} = (\overline{\underline{S}}_0, \overline{c}, \ldots)$ be two bicategories. A morphism $\Phi = (F, \varphi)$ from $\underline{S}$ to $\overline{\underline{S}}$ is determined by the following:

(i) A map $F: \underline{S}_0 \longrightarrow \overline{\underline{S}}_0$, $A \rightsquigarrow FA$.

(ii) A family of functors

$$F(A,B): \underline{S}(A,B) \longrightarrow \overline{\underline{S}}(FA,FB) \ , \quad S \rightsquigarrow FS \ , \quad s \rightsquigarrow Fs$$

(iii) For each object A of $\underline{S}$, an arrow of $\overline{\underline{S}}(FA,FA)$ (i.e. a 2-cell of $\overline{\underline{S}}$)

$$\varphi_A: \overline{I}_{FA} \longrightarrow F(I_A)$$

(iv) A family of natural transformations:

$$\varphi(A,B,C): \overline{c}(FA,FB,FC) \circ (F(A,B) \times F(B,C)) \longrightarrow F(A,C) \circ c(A,B,C).$$

$$\begin{array}{ccc}
\underline{S}(A,C) & \xleftarrow{\quad c(A,B,C) \quad} & \underline{S}(A,B) \times \underline{S}(B,C) \\
{\scriptstyle F(A,C)} \downarrow & \varphi(A,B,C) \ \nwarrow & \downarrow {\scriptstyle F(A,B) \times F(B,C)} \\
\overline{\underline{S}}(FA,FC) & \xleftarrow[\quad \overline{c}(FA,FB,FC) \quad]{} & \overline{\underline{S}}(FA,FB) \times \overline{\underline{S}}(FB,FC)
\end{array}$$

If (S,T) is an object of $\underline{S}(A,B) \times \underline{S}(B,C)$ the (S,T)-component of $\varphi(A,B,C)$

$$F(S \circ T) \xleftarrow{\varphi(A,B,C)(S,T)} FS \circ FT \quad (= FS \,\overline{\circ}\, FT) \qquad (*)$$

shall usually be abbreviated into $\varphi(S,T)$ or even φ.

(*) As usual in algebra, corresponding operations as c and $\overline{c}$ are in the abbreviated notation denoted by the same symbol, when no confusion is likely.

These data are required to satisfy the following coherence axioms:

(M. 1) If (S, T, U) is an object of $\underline{S}(A, B) \times \underline{S}(B, C) \times \underline{S}(C, D)$ the following diagram, where indices A, B, C, D have been omitted, is commutative:

$$\begin{array}{ccc}
FS\circ(FT\circ FU) & \xleftarrow[\sim]{\overline{a}(FS,FT,FU)} & (FS\circ FT)\circ FU \\
{\scriptstyle Id\circ\varphi(T,U)}\downarrow & & \downarrow{\scriptstyle \varphi(S,T)\circ Id} \\
FS\circ F(T\circ U) & & F(S\circ T)\circ FU \\
{\scriptstyle \varphi(S,T\circ U)}\downarrow & & \downarrow{\scriptstyle \varphi(S\circ T,U)} \\
F(S\circ(T\circ U)) & \xleftarrow[F(a(S,T,U))]{\sim} & F((S\circ T)\circ U)
\end{array}$$

(M. 2) If S is an object of $\underline{S}(A, B)$ the following diagrams commute:

$$\begin{array}{ccc}
FS & \xleftarrow[\sim]{Fr} & F(S\circ I_B) \\
{\scriptstyle \overline{r}}\uparrow\wr & & \uparrow{\scriptstyle \varphi(S,I_B)} \\
FS\circ\overline{I}_{FB} & \xrightarrow[Id\circ\varphi_B]{} & FS\circ FI_B
\end{array}
\qquad
\begin{array}{ccc}
F(I_A\circ S) & \xrightarrow[\sim]{F\ell} & FS \\
{\scriptstyle \varphi(I_A,S)}\uparrow & & \wr\uparrow{\scriptstyle \overline{\ell}} \\
FI_A\circ FS & \xleftarrow[\varphi_A\circ Id]{} & \overline{I}_{FA}\circ FS
\end{array}$$

(4. 2) Remark: The usual devices of universal algebra would have suggested the following "natural" notion of maps between bicategories $\underline{S}$ and $\overline{\underline{S}}$:

(i) A map $F: \underline{S}_o \longrightarrow \overline{\underline{S}}_o$, $A \rightsquigarrow FA$

(ii) A family of functors $F(A, B): \underline{S}(A, B) \longrightarrow \overline{\underline{S}}(FA, FB)$, commuting with the compositions; that is, $F(S\circ T) = FS\circ FT$ and $F(s\circ t) = F(s)\circ F(t)$, with the identities: $FI_A = \overline{I}_{FA}$, and with the a, ℓ, r :

$F(a(S, T, U)) = \overline{a}(FS, FT, FU)$, $F(\ell(S)) = \overline{\ell}(FS)$ and $F(r(S)) = \overline{r}(FS)$.

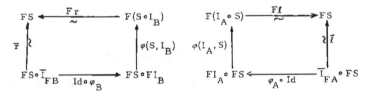

Such a map, called strict homomorphism can be, and will be, identified with the morphism $\Phi = (F,\varphi)$ defined by (i), (ii), and $\varphi_A = \mathrm{Id}: FI_A \xrightarrow{=} \bar{I}_{FA}$, and $\varphi(A,B,C)(S,T) = \mathrm{Id}: FS \circ FT \xrightarrow{=} F(S \circ T)$. We do not even require in (4.1) that F should commute up to isomorphisms with the compositions and units, i.e., that the φ_A and $\varphi(A,B,C)(S,T)$ should be isomorphisms. If this is satisfied, we say that $\Phi = (F,\varphi)$ is a <u>homomorphism.</u> If only the φ_A are isomorphisms, we say that Φ is a <u>unitary morphism</u>; if the φ_A are identities we say that Φ is a <u>strictly unitary morphism.</u>

The fact that all the desired results hold in the more general context, and, even more, the numerous mathematical examples where we have morphisms which are not homomorphisms, let alone strict ones, will be the essential justification of the definition (4.1). (See § 5)

<u>(4.3) Composition of morphisms.</u> Let $\underline{S} = (\underline{S}_o, c, \dots)$, $\bar{\underline{S}} = (\bar{\underline{S}}_o, \bar{c}, \dots)$ and $\bar{\bar{\underline{S}}} = (\bar{\bar{\underline{S}}}_o, \bar{\bar{c}}, \dots)$ be bicategories, $\Phi = (F,\varphi): \underline{S} \longrightarrow \bar{\underline{S}}$ and $\bar{\Phi} = (\bar{F},\bar{\varphi}): \bar{\underline{S}} \longrightarrow \bar{\bar{\underline{S}}}$ be morphisms: Construct the following

(i) A map $G = \bar{F} \circ F: \underline{S}_o \longrightarrow \bar{\bar{\underline{S}}}_o$.

(ii) A family of functors $G(A,B)$ as the composite:

$$\underline{S}(A,B) \xrightarrow{F(A,B)} \bar{\underline{S}}(FA,FB) \xrightarrow{\bar{F}(FA,FB)} \bar{\bar{\underline{S}}}(\bar{F}FA,\bar{F}FB) = \bar{\bar{\underline{S}}}(GA,GB)$$

(iii) For each object A of $\underline{S}$, an arrow ψ_A in $\underline{S}(GA,GA)$ as the composite:

$$\bar{\bar{I}}_{GA} = \bar{\bar{I}}_{\bar{F}FA} \xrightarrow{\bar{\varphi}_{FA}} \bar{F}\bar{I}_{FA} \xrightarrow{\bar{F}\varphi_A} \bar{F}FI_A = GI_A .$$

(iv) A family of natural transformations:

$$\psi(A,B,C): \bar{\bar{c}}(GA,GB,GC) \circ (G(A,B) \times G(B,C)) \longrightarrow G(A,C) \circ c(A,B,C)$$

by components $\psi(A,B,C)(S,T)$ for (S,T) objects of $\underline{S}(A,B) \times \underline{S}(B,C)$

making commutative the diagram:

$$\begin{array}{ccccc}
GS\circ GT & & \xrightarrow{\qquad \psi(A,B,C)(S,T) \qquad} & & G(S\circ T) \\
\Vert & & & & \Vert \\
\overline{F}FS\circ\overline{F}FT & \xrightarrow[\overline{\varphi}(FS,FT)]{} & \overline{F}(FS\circ FT) & \xrightarrow[\overline{F}(\varphi(S,T))]{} & \overline{F}F(S\circ T)
\end{array}$$

(The fact that the $\psi(A,B,C)(S,T)$ are natural follows from the 2-dimensional definition of $\psi(A,B,C)$ as the composite:

$$G\circ c = \overline{F}\circ F\circ c \xleftarrow{\overline{F}\circ\varphi} \overline{F}\circ\overline{c}\circ(F\times F) \xleftarrow{\overline{\varphi}\circ(F\times F)} \overline{\overline{c}}\circ(\overline{F}\times\overline{F})\circ(F\times F) = \overline{\overline{c}}\circ(G\times G)$$

where again indices A, B, C are omitted.)

<u>(4.3.1) Theorem.</u> With the previous notation

(i) The data $(G, G(A,B), \psi_A, \psi(A,B,C))$ define a morphism $(G,\psi) = \Psi$ from $\underline{S}$ to $\underline{\overline{\overline{S}}}$, called composite of Φ and $\overline{\Phi}$, and written $\overline{\Phi}\Phi$.

(ii) With this composition, bicategories and their morphisms form a category, which we will denote (*) $\mathrm{Bicat}^{[1]}$.

<u>Proof of (i).</u> To show that Ψ satisfies (M.1) we must prove that the exterior of the following diagram is commutative:

(*) Later on we will define a "trigraph" having $\mathrm{Bicat}^{[1]}$ as one-dimensional skeleton.

(Where S, T, U are objects of $\underline{S}(A, B)$, $\underline{S}(B, C)$, $\underline{S}(C, D)$)

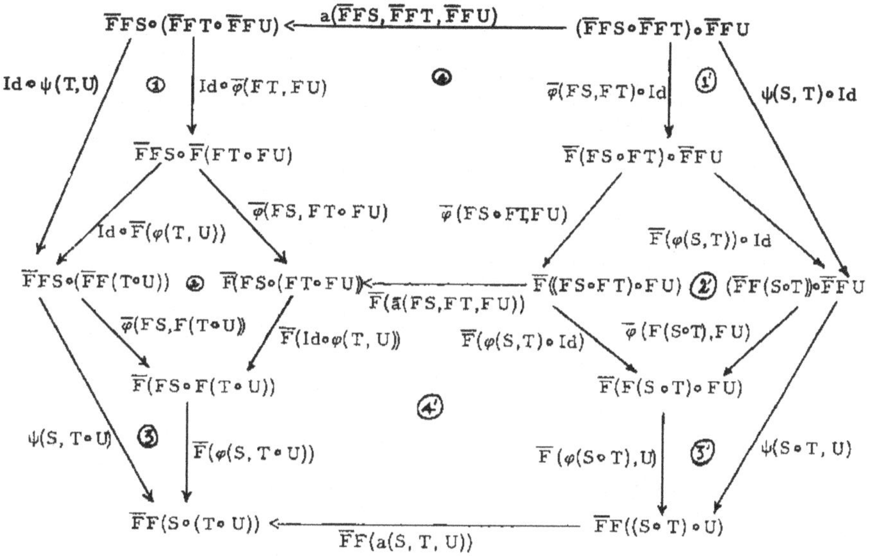

But the triangles 1 and 1' commute by definition of ψ and by the fact that $\overline{\overline{c}}$ is a bifunctor ; 3 and 3' commute by definition of ψ; 2 and 2' by naturality of $\overline{\varphi}$; 4 is the axiom (M. 1) for φ applied on the object (FS, FT, FU) and 4' is the image by the functor $\overline{F}(FA, FC)$ of the commutative diagram (M. 1) for φ.

To prove (M. 2) we have to show that for any object S of $\underline{S}(A, B)$ the exteriors of the following two diagrams commute:

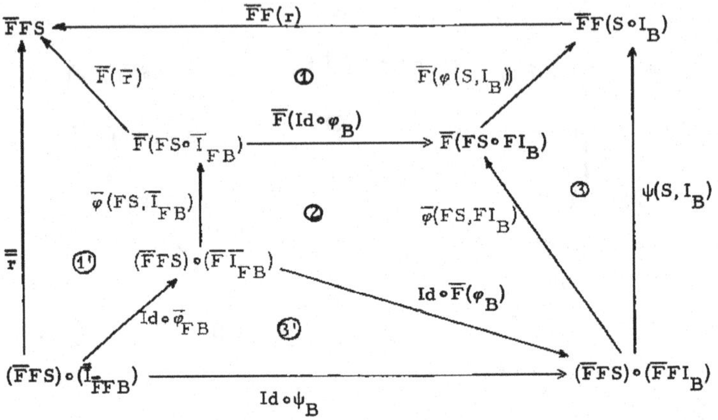

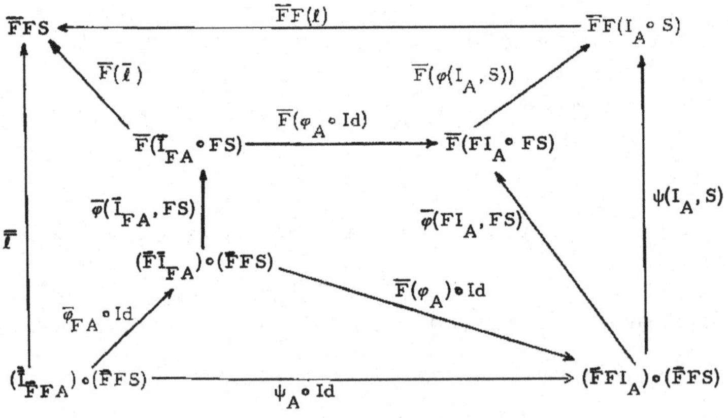

Now in the first diagram, the region 1 is the image by $\bar{F}$ of the diagram (M. 2) for r and $\bar{r}$, the region 1' is (M. 2) for $\bar{r}$ and $\bar{\bar{r}}$ applied on the object FS, then 2 commutes by naturality of φ, 3 by definition of $\psi(S, I_B)$ and 3' by definition of ψ_B and the fact that $\bar{\bar{c}}$ is a bifunctor. The commutativity of the second diagram can be proved similarly, or better, follows by transposition (cf. (3. 4. 4)) from the commutativity of the first.

Proof of (ii). Let $\bar{\bar{\Phi}} = (\bar{\bar{F}}, \bar{\bar{\varphi}}): \underline{\bar{S}} \longrightarrow \underline{\bar{\bar{S}}}$ be a third morphism. Denote by $\bar{\Psi} = (\bar{G}, \bar{\psi})$ the composite $\bar{\bar{\Phi}}\bar{\Phi}$, by $\Lambda = (L, \lambda)$ the composite $(\bar{\bar{\Phi}}\bar{\Phi})\Phi$ and by $(L', \lambda') = \Lambda'$ the composite $\bar{\bar{\Phi}}(\bar{\Phi}\Phi)$. The equations $L = L'$ and $L(A, B) = L'(A, B)$ for A, B objects of $\underline{S}_o$ are obvious. By definition of composition $\lambda_A = \lambda'_A$ is equivalent to the commutativity of the exterior of the diagram:

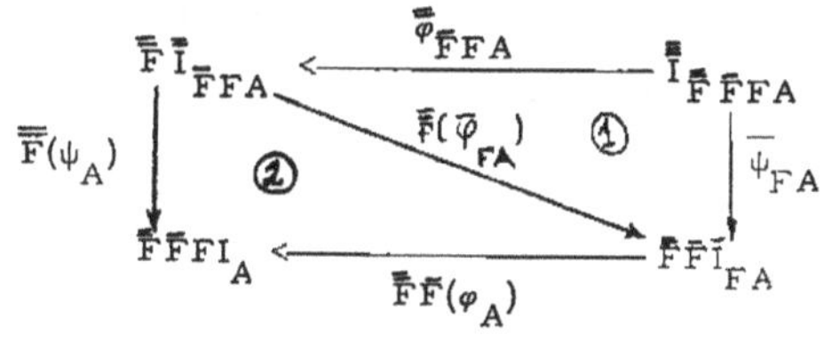

but 1 commutes by definition of $\bar{\psi}_{FA}$, and 2 is the image by $\bar{\bar{F}}$ of the commutative diagram defining ψ_A.

Similarly, for S, T objects of $\underline{S}(A, B)$, $\underline{S}(B, C)$, $\lambda(S, T) = \lambda'(S, T)$ is equivalent to the commutativity of the exterior of :

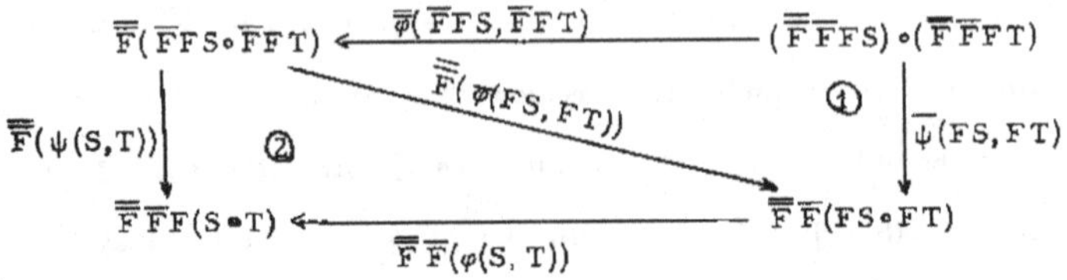

which follows again from the definitions of $\bar{\psi}(FS, FT)$ and $\psi(S, T)$. Thus composition of morphisms is associative. Finally, the data:

$Id: \underline{S}_0 \longrightarrow \underline{S}_0$, $Id(A, B) = Id: \underline{S}(A, B) \longrightarrow \underline{S}(A, B)$, $i_A = Id(I_A)$ and $i(A, B, C) = Id(c(A, B, C))$ obviously define a morphism $Id_{\underline{S}}: \underline{S} \longrightarrow \underline{S}$ which is an identity for the composition .

<u>(4.3.2) Proposition.</u> If Φ and $\bar{\Phi}$ are unitary, or strictly unitary, or homomorphisms or strict homomorphisms, so is their composite $\bar{\Phi}\Phi$.

Straightforward.

We shall denote the respective subcategories of $Bicat^{[1]}$ by $U\text{-}Bicat^{[1]}$, $S \circ U\text{-}Bicat^{[1]}$, $H\text{-}Bicat^{[1]}$ and $S \circ H\text{-}Bicat^{[1]}$.

<u>(4.3.3) Remark:</u> One can define, as in (1.3), morphisms of bicategories in a global diagrammatic way. This is left to the reader.

5. Examples

(5.1) Functors. Let $\underline{C}$ and $\overline{\underline{C}}$ be categories and $F: \underline{C} \longrightarrow \overline{\underline{C}}$ be a functor. Call $D\underline{C}$ and $D\overline{\underline{C}}$ the locally discrete bicategories associated with $\underline{C}$ and $\overline{\underline{C}}$. Then F obviously determines a morphism $DF: D\underline{C} \longrightarrow D\overline{\underline{C}}$. Conversely every morphism from $D\underline{C}$ to $D\overline{\underline{C}}$ comes from a unique such F, and obviously we get a full and faithful functor, called degeneracy

$$D: \mathrm{Cat}^{[1]} \longrightarrow \mathrm{Bicat}^{[1]}.$$

Moreover, DF is always a strict homomorphism, and more generally, if $\underline{S}$ is any bicategory and $\underline{C}$ a category, any morphism $\Phi: \underline{S} \longrightarrow D\underline{C}$ is a strict homomorphism. $D\underline{C}$ is called degenerate of $\underline{C}$.

(5.2) Multiplicative Categories. Let $\underline{M}$ and $\overline{\underline{M}}$ be c.m. and $\Phi = (F, \varphi, \lambda)$ a morphism from $\underline{M}$ to $\overline{\underline{M}}$ as defined in [B.1]. Call $\underline{S} = I\underline{M}$ and $\overline{\underline{S}} = I\overline{\underline{M}}$ the bicategories with a single object associated with $\underline{M}$ and $\overline{\underline{M}}$ in (2.2) and, with the same notations, define: a map $\hat{F}: \underline{S}_o \longrightarrow \overline{\underline{S}}_o$, $0 \rightsquigarrow \overline{0}$; a functor $\hat{F}(0,0) = F$, a map $\hat{\varphi}_o = \lambda$ and a natural transformation $\hat{\varphi}(0,0,0) = \varphi$. Then $(\hat{F}, \hat{\varphi})$ is a morphism $I\Phi: I\underline{M} \longrightarrow I\overline{\underline{M}}$. If $\overline{\Phi}: \overline{\underline{M}} \longrightarrow \overline{\overline{\underline{M}}}$ is another morphism of c.m., and $\overline{\Phi}\,\Phi$ is the composite (in the sense of [B.1]) we have $I(\overline{\Phi}\Phi) = (I\overline{\Phi})(I\Phi)$; and thus a functor:

$$I: \mathrm{Mult}^{[1]} \longrightarrow \mathrm{Bicat}^{[1]}.$$

It is clear that I is full and faithful, and that Φ is a homomorphism or a strict homomorphism of c.m. iff $I\Phi$ has the same property in $\text{Bicat}^{[1]}$. (This full and faithful embedding obviously extends to actions of c.m. on categories or to the situation considered by Epstein. cf (2.4)). If $\Phi^* = (F^*, \varphi^*, \lambda^*)$ is a comorphism of c.m. it can be, according to [B.1], identified with a morphism of the duals $\underline{M}^*$ and $\overline{\underline{M}}^*$, these in turn can be identified with the conjugates $(I\underline{M})^c$ and $(I\overline{\underline{M}})^c$ by (3.4.2). Thus the notion of comorphism is reduced, via a suitable duality to that of morphism.

(5.2.1) Remark: One should note that this identification is contravariant: the comorphisms from $\underline{M}$ to $\overline{\underline{M}}$ and the morphisms from $\underline{M}^*$ to $\overline{\underline{M}}^*$ can both be made, in a natural way, the objects of categories: $\text{Comor}(\underline{M}, \overline{\underline{M}})$ and $\text{Mor}(\underline{M}^*, \overline{\underline{M}}^*)$ dual to each other. Using the dualities of §3 one could define eight (!!!) different variances of morphisms between bicategories. The only way to avoid a cumbersome terminology is to consider always morphisms, and specify in each case the suitable dual categories for the domain and range.

(5.3) 2-Functors. Let $\underline{A}$ and $\overline{\underline{A}}$ be 2-categories and $F: \underline{A} \longrightarrow \overline{\underline{A}}$ a 2-functor as defined in [B.3]. Calling $J\underline{A}$ and $J\overline{\underline{A}}$ the strictly associative bicategories associated with $\underline{A}$ and $\overline{\underline{A}}$, F determines obviously a morphism $JF: J\underline{A} \longrightarrow J\overline{\underline{A}}$, and we get a functor:

$$J: \text{2-Cat}^{[1]} \longrightarrow \text{Bicat}^{[1]}$$

which is faithful but no longer full. Explicitely the morphisms from $J\underline{A}$ to $J\overline{\underline{A}}$ which are of form JF are exactly the strict homomorphisms.

In the rest of the paper we shall usually identify categories, c.m.'s and 2-categories with bicategories via the functors D, I, and J. All these examples have nothing surprising since the definition of bicategories was clearly devised to contain them. The following ones are of a completely different nature.

(5.4) Monads. Let $\underline{S}$ be a bicategory.

(5.4.1) Definition. A monad in $\underline{S}$ (or $\underline{S}$-monad) is a morphism from $\mathbb{1}$ to $\underline{S}$. An $\underline{S}$-comonad is an $\underline{S}^c$-monad.

Interpreting (4.1), a monad $\Phi = (F, \varphi): 1 \longrightarrow \underline{S}$ is determined by:

(i) One object $F(O) = X$ of $\underline{S}$; Φ is called an $\underline{S}$-monad on X or over X.

(ii) One functor $F(O,O): 1 \longrightarrow \underline{S}(X,X)$, i.e., an object T of $\underline{S}(X,X)$.

(iii) One arrow $\varphi_O = \eta : I_X \longrightarrow T$ in $\underline{S}(X,X)$.

(iv) One natural transformation $\varphi(O,O,O)$ identified with its unique component $\varphi(O,O,O)(O) = \mu : T \circ T \longrightarrow T$ in $\underline{S}(X,X)$.

The axiom (M.1) is equivalent to the commutativity of:

$$\begin{array}{ccc}
T\circ(T\circ T) & \xleftarrow[\sim]{a(T,T,T)} & (T\circ T)\circ T \\
{\scriptstyle T\circ\mu}\downarrow & & \downarrow{\scriptstyle \mu\circ T} \\
T\circ T & & T\circ T \\
{\scriptstyle \mu}\downarrow & & \downarrow{\scriptstyle \mu} \\
T & \xleftarrow[F(Id)\,=\,Id]{\sim} & T
\end{array}$$

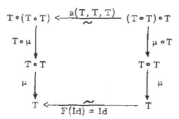

And (M. 2) to the commutativity of the diagrams:

$$\begin{array}{ccc} T & \xleftarrow[\sim]{Id} & T \\ {\scriptstyle r}\uparrow & & \uparrow{\scriptstyle \mu} \\ T\circ I_X & \xrightarrow[T\circ\eta]{} & T\circ T \end{array} \qquad \begin{array}{ccc} T & \xleftarrow[\sim]{Id} & T \\ {\scriptstyle \ell}\uparrow & & \uparrow{\scriptstyle \mu} \\ I_X\circ T & \xrightarrow[\eta\circ T]{} & T\circ T \end{array}$$

By suitably chosing $\underline{S}$ we will have many examples:

(5. 4. 1) Monoids: Take $\underline{S} = \underline{M} = (\underline{A}, \otimes, \dots)$ to be a c. m.; as a bicategory it has a unique object, say 0, thus X is determined. An $\underline{M}$-monad will therefore be defined by: an object T of $\underline{A}$, two arrows $\mu: T\otimes T \longrightarrow T$ and $\eta: \Lambda \longrightarrow T$. The commutativity of the previous diagrams is exactly the requirement that (T, μ, η) should be a monoid in $\underline{M}$, in the sense of [G]. (*) Dually, the $\underline{M}$-comonads are the co-monoids in $\underline{M}$. In particular, for $\underline{M} = (\text{Sets}, \times, \dots)$ we have the ordinary monoids, for $\underline{M} = (\underline{A}, \times, \dots)$ where $\underline{A}$ is a category with finite products, we have the monoid-like objects of [E. H], for further examples see [B. 2].

(5. 4. 2) Standard constructions: Take $\underline{S}$ = Cat, then X is a category, $T: X \longrightarrow X$ a functor, $\eta: Id_X \longrightarrow T$ and $\mu: TT \longrightarrow T$ natural transformations, a, ℓ, r are identities and the commutative diagrams express Godement's axioms for a standard construction, also called triple in [E. M]. By conjugation, Cat-comonads are identified with categories equipped with a co-construction or cotriple. We will usually abbreviate Cat-monads and Cat-comonads, to monads and comonads.

(*) Our choice of "monad" comes from this example and the definition (5. 5).

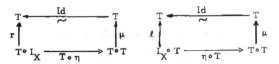

(5.4.3) Categories inside a category: Let $\underline{C}$ be any category with pull-backs $Sp\underline{C}$ the bicategory of spans of $\underline{C}$, of (2.6). We define a category inside $\underline{C}$ to be a monad of $Sp\underline{C}$.

Explicitly, such a category is defined by:

(i) An object A of $\underline{C}$, written X_o and called object of objects

(ii) A diagram T: $X_o \xleftarrow{d_1} X_1 \xrightarrow{d_o} X_o$; X_1 is the object of arrows, d_o and d_1 are called domain and codomain maps

(iii) A commutative diagram:

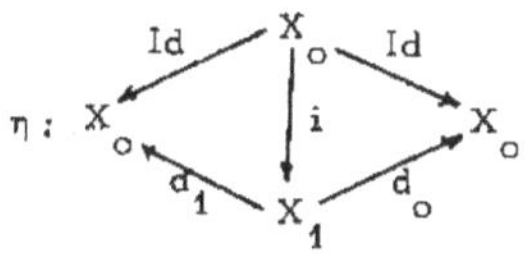

is thus determined by i which is called degeneracy or identity.

(iv) A commutative diagram

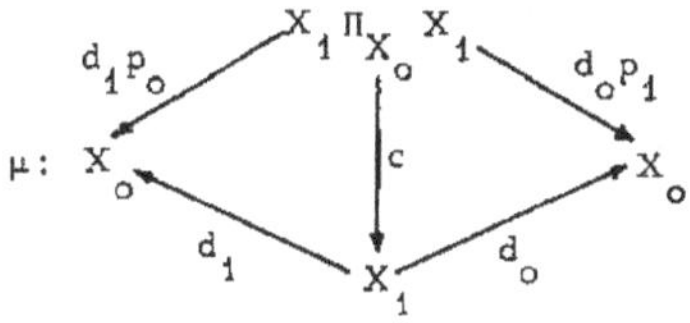

; p_o, p_1 defined by the pullback:

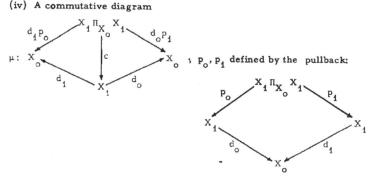

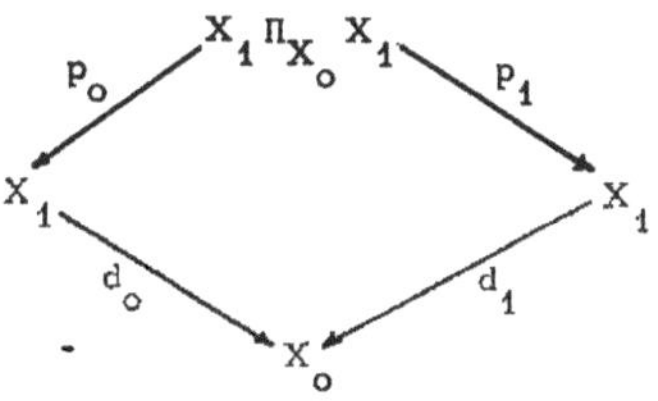

μ is determined by the previous maps and by c which is called multiplication or composition.

The maps (d_o, d_1, i, c) are required to make commutative three diagrams expressing the associativity of c and the fact that i is "an identity", which are left to the reader.

Taking $\underline{C}$ = Sets, we get the categories, $\underline{C} = \mathrm{Cat}^{[1]}$ we get the <u>double categories</u> of Ehresmann [Eh], $\underline{C}$ = Top = category of topological spaces we get the <u>topological categories</u>, etc. ...

Dually if $\underline{C}$ has pushouts, a <u>cocategory inside $\underline{C}$</u> is a comonad of Cosp $\underline{C}$. Explicitly, it is defined by two objects of $\underline{C}$, X_o and X_1 together with maps in $\underline{C}$:

$$X_o \underset{\partial_1}{\overset{\partial_o}{\rightrightarrows}} X_1 \xrightarrow{\sigma} X_o \qquad\qquad X_1 \xrightarrow{\gamma} X_1 \coprod_{X_o} X_1$$

satisfying "well-known" axioms, which can best be visualized by looking, inside $\mathrm{Cat}^{[1]}$, at the <u>fundamental cocategory</u> described as follows: $X_o = \mathbf{1}$, $X_1 = \mathbf{2}$: $0 \longrightarrow 1$, $\sigma, \partial_o, \partial_1$ are the only possible distinct functors. Then $X_1 \coprod_{X_o} X_1$ is $\mathbf{3}$, i.e. $0 \longrightarrow 1 \longrightarrow 2$ and γ is the functor sending the non degenerate arrow of $\mathbf{2}$ on $0 \longrightarrow 2$ in $\mathbf{3}$.

<u>(5.4.4) Ordered sets:</u> If $\underline{S}$ is a locally ordered bicategory, in any of the categories $\underline{S}(A, B)$ all diagrams commute; thus a monad in $\underline{S}$ is determined by an object A of $\underline{S}$, an object T of $\underline{S}(A, A)$ such that $I_A \leq T$ and $T \circ T \leq T$ with no further conditions. In particular, in the bicategory $\underline{R}$ of relations, A is a set, T a subset of $A \times A$ such that $\Delta_A \subseteq T$ and $T \circ T \subset T$. Thus the monads of $\underline{R}$ are the partially ordered sets.

Note that all the examples of (5.4) would have been reduced to objects of the different bicategories $\underline{S}$ involved, had we confined ourselves to strict homomorphisms, since the domain was $\mathbb{1}$.

(5.5) Polyads. We now consider morphisms with domain slightly more general, namely a locally punctual bicategory, that is according to (2.7) a bicategory $\underline{S}$ such that $\underline{S}(A, B) = \mathbb{1}$ for all objects A, B. Such a bicategory is clearly determined by the set $\operatorname{Ob}\underline{S}$ of its objects.

(5.5.1) Definition. Let $\overline{\underline{S}}$ be a bicategory. A polyad in $\overline{\underline{S}}$ (or $\overline{\underline{S}}$-polyad) is a morphism of bicategories $\Phi = (F, \varphi): \underline{S} \longrightarrow \overline{\underline{S}}$ where $\underline{S}$ is locally punctual. The set $\operatorname{Ob}\underline{S}$ is called set of objects or indices of the polyad. (The monads are obtained when $\operatorname{Ob}\underline{S} = \mathbb{1}$, hence the name of polyad.)

We will give three examples. By suitably choosing $\overline{\underline{S}}$ in the list of examples of §2, the reader can construct many more.

(5.5.2) Relative categories: Let $\underline{M} = (\underline{A}, \otimes, \Lambda, \ldots)$ be a multiplicative category. Let us recall our definition [B.3] of an $\underline{M}$-category $\underline{C}$. It is given by

(1) A set $\operatorname{Ob}(\underline{C})$ whose elements $X, Y, \ldots$ are called objects of $\underline{C}$.

(2) For each X, Y in $\operatorname{Ob}(\underline{C})$ an object $\underline{C}(X, Y)$ of $\underline{A}$.

(3) For each X, Y, Z in $\operatorname{Ob}(\underline{C})$ a map of $\underline{A}$, $c(X, Y, Z)$ abbreviated in c,

$$c = c(X, Y, Z) : \underline{C}(X, Y) \otimes \underline{C}(Y, Z) \longrightarrow \underline{C}(X, Z)$$

(4) For each $X \in \operatorname{Ob}\underline{C}$, a map of $\underline{A}$, $i_{\underline{C}}(X)$ abbreviated in $i(X)$ or i,

$$i = i(X): \Lambda \longrightarrow \underline{C}(X, X)$$

such that, for all X, Y, Z, T in Ob($\underline{C}$) the following diagrams commute:

$$\begin{array}{ccccc}
(\underline{C}(X,Y)\otimes\underline{C}(Y,Z))\otimes\underline{C}(Z,T) & \xrightarrow{c\otimes Id} & \underline{C}(X,Z)\otimes\underline{C}(Z,T) & & \\
 & & & \searrow^{c} & \\
\theta\Big\downarrow\wr & & & & \underline{C}(X,T) \\
 & & & \nearrow_{c} & \\
\underline{C}(X,Y)\otimes(\underline{C}(Y,Z)\otimes\underline{C}(Z,T)) & \xrightarrow{Id\otimes c} & \underline{C}(X,Y)\otimes\underline{C}(Y,T) & &
\end{array}$$

$$\begin{array}{ccccc}
 & & \underline{C}(X,X)\otimes\underline{C}(X,Y) & & \\
 & {}^{i\otimes Id}\nearrow & & \searrow^{c} & \\
\Lambda\otimes\underline{C}(X,Y) & & \xrightarrow[\gamma]{\sim} & & \underline{C}(X,Y)
\end{array}
\qquad
\begin{array}{ccccc}
 & & \underline{C}(X,Y)\otimes\underline{C}(Y,Y) & & \\
 & {}^{Id\otimes i}\nearrow & & \searrow^{c} & \\
\underline{C}(X,Y)\otimes\Lambda & & \xrightarrow[\delta]{\sim} & & \underline{C}(X,Y)
\end{array}$$

For examples we refer to [B. 3].

Now if $\underline{C}$ is such an $\underline{M}$-category, take $\underline{S}$ to be the locally punctual bicategory having Ob($\underline{C}$) as set of objects, and $\overline{\underline{S}} = I\underline{M}$ to be the bicategory with one object $\mathbb{O}$ (cf (5. 2)), and define:

(i) A map $F: \underline{S}_o \longrightarrow \overline{\underline{S}}_o$ as the unique map $Ob(\underline{C}) \longrightarrow \mathbf{1}$

(ii) Functors $F(X,Y)$: $\underline{S}(X,Y) = \mathbf{1} \longrightarrow \overline{\underline{S}}(FX,FY) = \underline{A}$ by

$$F(X,Y)(\mathbb{O}) = \underline{C}(X,Y).$$

(iii) Arrows $\varphi_X: \overline{I}_{FX} = \Lambda \longrightarrow F(I_X) = F(X,X)(\mathbb{O}) = \underline{C}(X,X)$, by $\varphi_X = i_X$.

(iv) Natural transformations $\varphi(X,Y,Z)$ identified to their unique component $\varphi(X,Y,Z)(\mathbb{O},\mathbb{O})$ by

$$\varphi(X,Y,Z) = c(X,Y,Z): F(X,Y)(\mathbb{O})\otimes F(Y,Z)(\mathbb{O}) = \underline{C}(X,Y)\otimes\underline{C}(Y,Z) \longrightarrow \underline{C}(X,Z)$$
$$= F(X,Z)(\mathbb{O}).$$

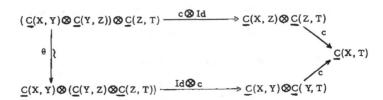

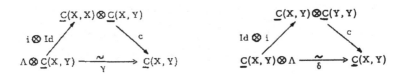

One easily verifies that the commutativity of the previous diagrams is then equivalent to (M.1) and (M.2). Thus (F, φ) is a morphism $\Phi(\underline{C}) : \underline{S} \longrightarrow \overline{\underline{S}}$.

Conversely, given a polyad $\Phi: \underline{S} \longrightarrow \overline{\underline{S}}$ where $\overline{\underline{S}}$ has a single object $\mathbf{O}$, one defines a category $\underline{C}(\Phi)$ relative to the c.m. $\underline{S}(\mathbf{O}, \mathbf{O})$, having $\mathrm{Ob}(\underline{S})$ as set of objects, in an obvious manner.

(5.5.3) Proposition. The assignments $\underline{C} \rightsquigarrow \Phi(\underline{C})$ and $\Phi \rightsquigarrow \underline{C}(\Phi)$ establish a bijection between categories relative to multiplicative categories and morphisms of bicategories with domain locally punctual and codomain having a single object.

(5.5.4) Coherent families of isomorphisms. Frequently looking for objects X of a category $\underline{C}$ having some properties (e.g. universal properties, or objects obtained by iteration of a tensor product associative up to isomorphisms) one finds a whole family of such X_i, indexed by a set I, equipped with "canonical" isomorphisms $\varphi_{ij}: X_i \longleftarrow X_j$, such that $\varphi_{ii} = \mathrm{Id}$, $\varphi_{ij}\varphi_{jk} = \varphi_{ik}$. Such families can be obviously identified with $\underline{C}$-polyads indexed by I (where $\underline{C}$ is as usual identified with the degenerate bicategory $D\underline{C}$).

(5.5.5) Polyspans. Let $\underline{C}$ be a category with pullbacks, $\mathrm{Sp}\underline{C}$ the bicategory of spans of $\underline{C}$. A polyad in $\mathrm{Sp}\underline{C}$, indexed by a set I, is determined by:

(i) A map $F: I \longrightarrow \mathrm{Ob}\,\mathrm{Sp}\,\underline{C} = \mathrm{Ob}\,\underline{C}$, written $i \rightsquigarrow X_i$.

(ii) For each pair (i, j), a functor $F(i, j): \mathbb{1} \longrightarrow \mathrm{Sp}\,\underline{C}(X_i, X_j)$, identified with an object S_{ij} of $\mathrm{Sp}\,\underline{C}(X_i, X_j)$, that is a diagram in $\underline{C}$:

$$S_{ij}: X_i \xleftarrow{\;g_{ij}\;} X_{ij} \xrightarrow{\;\overline{g}_{ij}\;} X_j .$$

(iii) For each $i \in I$, an arrow $\varphi_i: I_{X_i} \longrightarrow S_{ii}$ in $\mathrm{Sp}(X_i, X_i)$, that is a commutative diagram in $\underline{C}$:

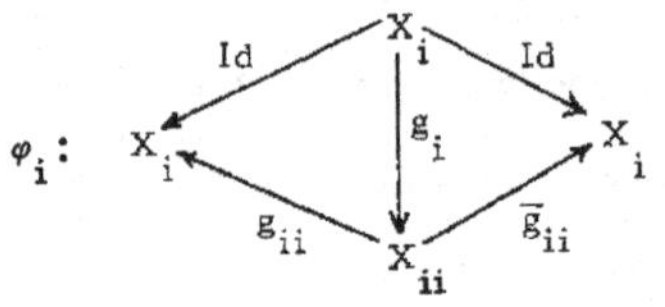

(iv) For each (i, j, k) a natural transformation $\varphi(i, j, k)$ determined by its unique component $\varphi(i, j, k)(\mathbb{O}, \mathbb{O}): S_{ij} \circ S_{jk} \longrightarrow S_{ik}$, that is according to (2.6) a commutative diagram in $\underline{C}$; where p and $\overline{p}$ are the projections of the pullback:

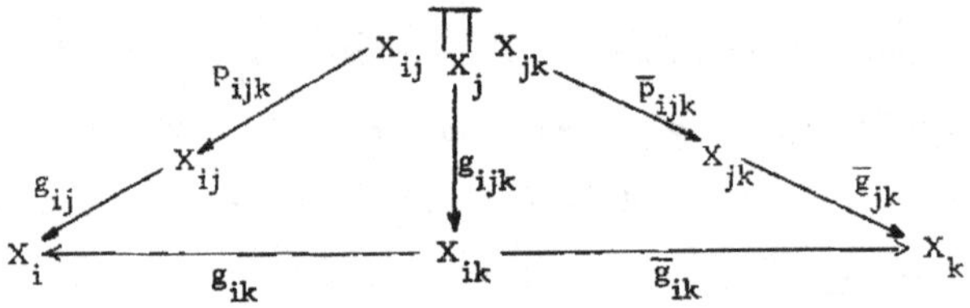

Note that all these data are determined by the maps $g_i, g_{ij}, \overline{g}_{ij}, g_{ijk}$. The conditions (M. 1) and (M. 2) are expressed, in terms of these maps, by the commutativity of the three diagrams below, where the notation $X_{ij} \bullet X_{jk}$

stands for $X_{ij} \prod_{X_j} X_{jk}$, $X_{ij} \circ X_j$ for $X_{ij} \prod_{X_j} X_j$, and a, ℓ, r are the isomorphisms of associativity and identity of pullbacks.

$$\begin{array}{ccc}
(X_{ij} \circ X_{jk}) \circ X_{k\ell} & \xrightarrow[\sim]{a} & X_{ij} \circ (X_{jk} \circ X_{k\ell}) \\
{\scriptstyle g_{ijk} \circ Id_{k\ell}} \downarrow & & \downarrow {\scriptstyle Id_{ij} \circ g_{jk\ell}} \\
X_{ik} \circ X_{k\ell} & & X_{ij} \circ X_{j\ell} \\
{\scriptstyle g_{ik\ell}} \downarrow & & \downarrow {\scriptstyle g_{ij\ell}} \\
X_{i\ell} & \xrightarrow{Id_{i\ell}} & X_{i\ell}
\end{array}$$

And:

$$\begin{array}{ccc}
X_{ij} \circ X_j & \xrightarrow{Id_{ij} \circ g_{jj}} & X_{ij} \circ X_{jj} \\
{\scriptstyle r} \downarrow \wr & & \downarrow {\scriptstyle g_{ijj}} \\
X_{ij} & \xleftarrow{Id_{ij}} & X_{ij}
\end{array}
\qquad
\begin{array}{ccc}
X_i \circ X_{ij} & \xrightarrow{g_{ii} \circ Id_{ij}} & X_{ii} \circ X_{ij} \\
{\scriptstyle \ell} \downarrow \wr & & \downarrow {\scriptstyle g_{iij}} \\
X_{ij} & \xleftarrow{Id_{ij}} & X_{ij}
\end{array}$$

That is, neglecting the a, ℓ, r which is always possible according to Theorem (B. 4), the <u>polyspans</u> of $\underline{C}$ satisfy the cocycle conditions:

(P. 1) $\quad g_{ik\ell}(g_{ijk} \circ Id_{k\ell}) = g_{ij\ell}(Id_{ij} \circ g_{jk\ell})$

(P. 2) $\quad g_{ijj}(Id_{ij} \circ g_{jj}) = Id \quad , \quad g_{iij}(g_{ii} \circ Id_{ij}) = Id.$

The significance of these equations in descent theory and non-abelian cohomology shall be examined elsewhere.

<u>(5. 6) Pseudo-functors.</u> In [Gr], Grothendieck defines a <u>pseudo-functor</u> $\underline{E}^* \longrightarrow Cat$, where $\underline{E}$ is a category, as:

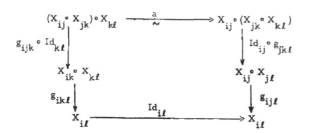

(a) A map $S \rightsquigarrow \underline{F}_S$ from $\mathrm{Ob}\,\underline{E}$ to Cat.

(b) For each $f: T \longrightarrow S$ in $\underline{E}$, a functor $f^*: \underline{F}_S \longrightarrow \underline{F}_T$.

(c) For each pair (f, g) of maps of $\underline{E}$ such that fg is defined, a natural transformation $c_{f,g}: g^*f^* \longrightarrow (fg)^*$.

(d) For each object S of $\underline{E}$, a natural transformation $\alpha_S: (\mathrm{Id}_S)^* \longrightarrow \mathrm{Id}_{\underline{F}_S}$.

These data are required to satisfy:

$$(A) \quad \begin{cases} c_{f,\mathrm{id}_T}(\xi) = \alpha_T(f^*(\xi)) \\ c_{\mathrm{id}_S,f}(\xi) = f^*(\alpha_S(\xi)) \end{cases}$$

$$(B) \qquad c_{f,gh}(\xi) \circ c_{g,h}(f^*(\xi)) = c_{fg,h}(\xi) \circ h^*(c_{f,g}(\xi))$$

for any maps $f: T \longrightarrow S$, $g: U \longrightarrow T$, $h: V \longrightarrow U$ in $\underline{E}$, and object ξ of $\underline{F}_S$.

He also considers the following special cases:

(i) For all S, $(\mathrm{id}_S)^* = \mathrm{id}_{\underline{F}_S}$ and the α_S are identities. (A) reduces to:

$$(A') \qquad c_{f,\mathrm{id}_T} = \mathrm{id}_{f^*}, \quad c_{\mathrm{id}_S,f} = \mathrm{id}_{f^*}$$

which he calls normalized.

(ii) All the $c_{f,g}$ are isomorphisms (this corresponds to fibered categories).

(iii) For all f, g, $(fg)^* = g^*f^*$ and $c_{fg} = \mathrm{Id}$ (this corresponds to split-fibrations, or functors $\underline{E}^* \longrightarrow \mathrm{Cat}$).

Define, given such a pseudo-functor $\underline{P} = (\underline{F}, *, c, \alpha)$, the following:

(1) A map $F: \mathrm{Ob}\,D\underline{E}^* = \mathrm{Ob}\,\underline{E} \longrightarrow \mathrm{Ob}\,\mathrm{Cat}$, by $FS = \underline{F}_S$. (Where $D\underline{E}^*$ is the degenerate bicategory defined by $\underline{E}^*$.)

(2) If f is an object of the discrete category $D\underline{E}^*(T,S)$, $F(T,S)(f) = f^*$.

(3) $\varphi(U,T,S)$ to be the natural transformation having as components the natural transformations $\varphi(U,T,S)(g,f) = c_{f,g}$.

(4) Arrows φ_S in $\mathrm{Cat}(\underline{F}_S, \underline{F}_S)$ to be the natural transformation α_S.

(5.6.1) Theorem: With the previous notations:

(a) $(F, F(S,T), \varphi_S, \varphi(U,T,S))$ define a morphism $\Phi(\underline{P})$: $D\underline{E}^* \longrightarrow \mathrm{Cat}$.

(b) The correspondence $\underline{P} \rightsquigarrow \Phi(\underline{P})$ is a bijection between pseudo-functors and morphisms of bicategories with domain a category and co-domain Cat.

(c) Under this correspondence the pseudo-functors satisfying (i), (ii), and (iii) become respectively the strictly unitary morphisms, the homomorphisms and the strict homomorphisms.

The proof is a straightforward and tedious verification that the requirements (A) and (B) for pseudo-functors, are equivalent in this case to (M. 2) and (M. 1) of (4. 11) respectively, and then (c) is a rephrasing of the definitions.

In Part II, the construction of [Gr] assigning to each morphism, i. e., pseudo-functor, $\underline{E}^* \longrightarrow \mathrm{Cat}$ a category $\underline{F}$ equipped with a functor $p: \underline{F} \longrightarrow \underline{E}$ together with a cleavage of p, shall be extended by replacing Cat by the bigger bicategory Prof of profunctors. Then all the properties of categories over a category $\underline{E}$ -- fibrations, cofibrations, cleavages, splittings,... -- will have simple interpretations in terms of morphisms $\underline{E}^* \longrightarrow \mathrm{Prof}$. We will also extend the construction to the case where the domain is any bicategory, not necessarily one-dimensional.

<u>(5.7) Bimodules and Rings.</u> Let Ring be the category of rings with identity, Bim the bicategory defined in (2.6). With the same notations, define:

(i) A map $F = Id : Ob(Ring) \longrightarrow Ob(Bim)$.

(ii) Functors $F(A,B)$: $Ring(A,B) \longrightarrow Bim(A,B) = {}_A\underline{M}_B$, $f \rightsquigarrow M_f$.

(iii) For each A, a map of bimodules $\varphi_A = Id$: $A \longrightarrow A$.

(iv) Natural transformations $\varphi(A,B,C)$ by their components:

$$\varphi(A,B,C)(f,g): \; M_f \circ M_g \xrightarrow{\sim} M_{fg} \, .$$

Then (F,φ): Ring $\longrightarrow$ Bim is a homomorphism, strictly unitary. (Here , of course, Ring has been identified with the degenerate bicategory D Ring .) Moreover, the correspondence $A \rightsquigarrow A$, $f \rightsquigarrow M_f$ embeds the category of rings in the bicategory of bimodules.

<u>(5.8) Functoriality of spans and cospans.</u> Let $\underline{C}$ and $\overline{\underline{C}}$ be any categories with pushouts and F: $\underline{C} \longrightarrow \overline{\underline{C}}$ be a functor. Choose pushouts in $\underline{C}$ and $\overline{\underline{C}}$, and define:

(i) A map F: $Ob(Cosp\,\underline{C}) = Ob(\underline{C}) \longrightarrow Ob(Cosp\,\overline{\underline{C}}) = Ob(\overline{\underline{C}})$, $X \rightsquigarrow FX$.

(ii) Functors $F(A,B)$: $Cosp\,\underline{C}(A,B) \longrightarrow Cosp\,\overline{\underline{C}}(FA,FB)$ by:

$$(S: A \xrightarrow{\alpha} X \xleftarrow{\beta} B) \rightsquigarrow (FS: FA \xrightarrow{F\alpha} FX \xleftarrow{F\beta} FB).$$

(iii) For each $A \in Ob\,\underline{C}$ an arrow $\varphi_A = Id$: $I_{FA} \longrightarrow F_{IA}$ of $Cosp\,\underline{C}(FA,FA)$

(iv) If $S: A \longrightarrow X \longleftarrow B$ and $T: B \longrightarrow Y \longleftarrow C$, a map

$\varphi(A,B,C)(S,T)$: $FS \circ FT \longrightarrow F(S \circ T)$ in $Cosp\,\overline{\underline{C}}(FA,FC)$ to be the diagram:

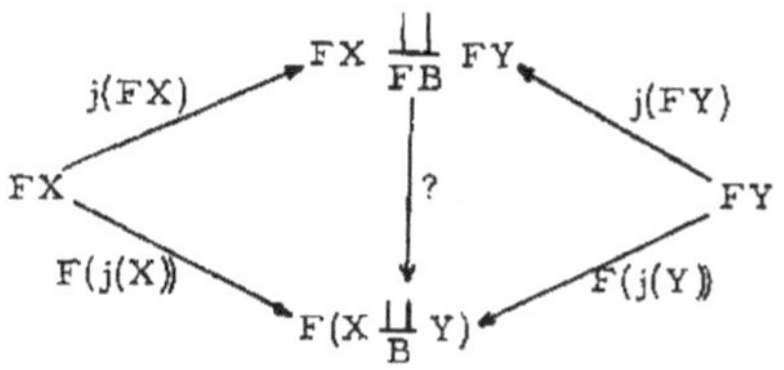

where $j(FX), j(FY), j(X), j(Y)$ are the canonical maps in the pushouts and ? is the unique map making the diagram commutative (there is always one such).

(5.8.1) Proposition. With the previous notations:

(i) (F, φ) determine a strictly unitary morphism $\mathrm{Cosp}F$: $\mathrm{Cosp}\underline{C} \longrightarrow \mathrm{Cosp}\overline{\underline{C}}$.

(ii) $\mathrm{Cosp}F$ is a homomorphism iff F commutes with pushouts.

(iii) $\mathrm{Cosp}F$ is a strict homomorphism iff F commutes with the chosen pushouts.

(iv) If $\overline{F}: \overline{\underline{C}} \longrightarrow \overline{\overline{\underline{C}}}$ is another functor, choosing pushouts in $\overline{\overline{\underline{C}}}$ we get $\mathrm{Cosp}(\overline{F}F) = \mathrm{Cosp}(\overline{F})\,\mathrm{Cosp}(F)$.

Proof long but straightforward.

Note that if $\underline{C}$ and $\overline{\underline{C}}$ have pullbacks and $F: \underline{C} \longrightarrow \overline{\underline{C}}$ is a functor, $\mathrm{Sp}F$ defined dually is a comorphism from $\mathrm{Sp}\underline{C}$ to $\mathrm{Sp}\overline{\underline{C}}$ (i.e., a morphism of the conjugates). In particular, if $\underline{C}$ and $\overline{\underline{C}}$ have final objects 1 and $\overline{1}$ any functor defines a comorphism of the multiplicative categories $\underline{C}$ and $\overline{\underline{C}}$ (with $\prod$ as multiplication), since $\underline{C}$ and $\overline{\underline{C}}$ are equivalent to the c.m.'s $\mathrm{Sp}\underline{C}(1,1)$ and $\mathrm{Sp}\overline{\underline{C}}(\overline{1},\overline{1})$ by restrictions of $\mathrm{Sp}F$. It will be a homomorphism (resp. strict) of c.m. iff F commutes with products (resp. with chosen products).

§6. Some Corollaries of Theorem (4.3.1):

The interest of defining mathematical objects as morphisms of bi-categories, is the possibility to compose them with other morphisms to get new objects as direct or inverse images. We give a few instances, many others can be obtained by choosing a pair of composable morphisms in the list of §5.

(6.1) Proposition: Let $\underline{M}$ and $\overline{\underline{M}}$ be multiplicative categories, $\Phi = (F, \varphi, \lambda)$: $\underline{M} \longrightarrow \overline{\underline{M}}$ a morphism, and $M = (T; \mu; \eta)$ a monoid of $\underline{M}$; then $(FT; F\mu \circ \varphi(T,T); F\eta \circ \lambda)$ is a monoid of $\overline{\underline{M}}$ called image of M by Φ, and denoted $\Phi(M)$.

Identify M with a morphism $\overline{\Phi} : \mathbb{1} \longrightarrow \underline{M}$, then $\Phi(M)$ is identified with $\Phi\overline{\Phi}$.

As an example, take (i) $\underline{M} = (\underline{A}, \otimes, \ldots)$, $\overline{\underline{M}}$ = the category of endomorphisms of $\underline{A}$ and Φ the left representation [B.1]; $A \rightsquigarrow A \otimes -$. To each monoid in $\underline{M}$ corresponds a monad over $\underline{A}$.
(ii) $\underline{M}$ and $\overline{\underline{M}}$ to be the endomorphisms of two categories $\underline{K}$ and $\underline{L}$. S: $\underline{K} \longrightarrow \underline{L}$ and T: $\underline{L} \longrightarrow \underline{K}$ a pair of adjoint functors, and Φ: $\underline{M} \longrightarrow \overline{\underline{M}}$ the morphism determined by the adjunction (cf [B.1]), to each monad on $\underline{K}$ corresponds a monad on $\underline{L}$.

(6.2). If $\underline{C}$ is a category with pushouts, $C = (X_0, X_1, \partial_0, \partial_1, \sigma, \gamma)$ a cocategory inside $\underline{C}$, cf (5.4.3) and F: $\underline{C}^* \longrightarrow \overline{\underline{C}}$ a functor which commutes with pullbacks, then $(FX_0, FX_1, \ldots)$ define a category FC inside $\overline{\underline{C}}$. Identify C with a morphism $\mathbf{1} \longrightarrow \mathrm{Sp}\underline{C}^*$ and note that F determines a morphism $\mathrm{Sp}\underline{C}^* \longrightarrow \mathrm{Sp}\overline{\underline{C}}$.

In particular, for each object X of $\underline{C}$, $\mathrm{Hom}(C,X)$ is a category, and $X \rightsquigarrow \mathrm{Hom}(C,X)$ a functor $\mathrm{Hom}(C,-)\colon \underline{C} \longrightarrow \mathrm{Cat}^{[1]}$. Taking $\underline{C} = \mathrm{Cat}^{[1]}$ and $C = \mathbf{2}$, we find that for each category $\underline{X}$, $\mathrm{Cat}^{[1]}(\mathbf{2}, \underline{X})$ is a category. The structure of Cat as a 2-category comes from this remark which will be generalized to get the 2 and 3 dimensional parts of Bicat.

(6.3). Let $\underline{M}$ be a c.m., $\underline{C}$ be an $\underline{M}$ category (cf (5.5.2)), $\underline{C}'_o$ a set and $f\colon \underline{C}'_o \longrightarrow \mathrm{Ob}(\underline{C})$ a map. For all X', Y', Z' in $\underline{C}'_o$ define

$$\underline{C}'(X',Y') = \underline{C}(fX',fY') \;,\quad i_{\underline{C}'}(X') = i_{\underline{C}}(fX') \;,\quad c'(X',Y',Z') = c(fX',fY',fZ').$$

(6.3.1) Proposition. With these notations, $(\underline{C}'_o, \underline{C}'(X,Y), i_{\underline{C}'}, c')$ is an $\underline{M}$ category $f^*(\underline{C})$ called inverse image of $\underline{C}$ by f, and the inverse images are transitive (i.e., $g^*f^*(\underline{C}) = (fg)^*(\underline{C})$; $\mathrm{id}^*(\underline{C}) = \underline{C}$).

Let $\underline{L}$ and $\underline{L}'$ be the locally punctual bicategories having $\mathrm{Ob}\,\underline{C}$ and $\underline{C}'_o$ as set of objects, $\Phi'\colon \underline{L}' \longrightarrow \underline{L}$ the morphism obviously determined by f and $\Psi\colon \underline{L} \longrightarrow I\underline{M}$ the morphism identified with $\underline{C}$ (cf (5.5.3)). Then $f^*(\underline{C})$ is the $\underline{M}$-category identified with the morphism $\Psi\Phi'\colon \underline{L}' \longrightarrow I\underline{M}$.

Let $\overline{\underline{M}}$ be another c.m. and $\Phi = (F, \varphi, \lambda)\colon \underline{M} \longrightarrow \overline{\underline{M}}$ a morphism. For all X, Y, Z in $\mathrm{Ob}(\underline{C})$ define: $\overline{\underline{C}}(X,Y) = F(\underline{C}(X,Y))$, $\overline{c}(X,Y,Z)$ to be the composite morphism:

$$F(\underline{C}(X,Y))\overline{\otimes}F(\underline{C}(Y,Z)) \xrightarrow{\varphi} F(\underline{C}(X,Y)\otimes\underline{C}(Y,Z)) \xrightarrow{F(c)} F(\underline{C}(X,Z))$$

and $i_{\overline{\underline{C}}}(X)$ to be the composite:

$$\overline{\Lambda} \xrightarrow{\lambda} F(\Lambda) \xrightarrow{F(i_{\underline{C}}(X))} F(\underline{C}(X,X)) = \overline{\underline{C}}(X,X).$$

(6.3.2) Proposition: With these notations $(Ob(\underline{C}), \overline{\underline{C}}(X,Y), i_{\overline{\underline{C}}}, \tau)$ is an $\overline{\underline{M}}$-category, $\Phi_*(\underline{C})$ called direct image of $\underline{C}$ by Φ, and direct images are transitive.

If $\underline{C}$ is identified with $\Psi\colon \underline{L} \longrightarrow I\underline{M}$, then $\Phi_*(\underline{C})$ is identified with the composite $I\Phi\circ\Psi\colon \underline{L} \longrightarrow I\underline{M} \longrightarrow I\overline{\underline{M}}$.

Note that, moreover, from the associativity of composition of morphisms, it follows that direct and inverse image commute with each other, that is: $\Phi_*(f^*(\underline{C})) = f^*(\Phi_*(\underline{C}))$.

(6.4). Let P be a pseudo-functor from $\underline{E}^*$ to Cat (cf. (5.6)), and $g\colon \overline{\underline{E}} \longrightarrow \underline{E}$ be a functor. P can be identified with a morphism $\Phi\colon D(\underline{E}^*) \longrightarrow$ Cat, g determines a morphism $D(g^*)\colon D(\overline{\underline{E}}^*) \longrightarrow D(\underline{E}^*)$ and the composite $\Phi\circ D(g^*)\colon D(\overline{\underline{E}}^*) \longrightarrow$ Cat defines a pseudo-functor $g^*(P)$ called inverse image of P by g ; again transitive. If $p\colon \underline{F} \longrightarrow \underline{E}$ is the cleaved category over $\underline{E}$ associated with P, then the cleaved category associated with $g^*(P)$ is the inverse image $g^*(p)$ (in the sense of cleaved categories i.e., the pullback $\overline{\underline{E}} \times \underline{F}$ with the cleavage pulled back from $\underline{F}$).

§7. Some Basic Constructions.

(7.1) Poincaré category. Let $\underline{S}$ be a bicategory. For each pair (A, B) of objects of $\underline{S}$, let $\Pi\underline{S}(A, B)$ be the set of connected components of the category $\underline{S}(A, B)$. If S is an object of $\underline{S}(A, B)$ we write $[S]_\Pi$ for its equivalence class. We define composition:

$$\Pi\underline{S}(A, B) \times \Pi\underline{S}(B, C) \longrightarrow \Pi\underline{S}(A, C) \text{ by } [S]_\Pi \circ [T]_\Pi = [S \circ T]_\Pi .$$

It is well-defined, associative and the $[I_A]_\Pi$ are identities, giving rise to a category ΠS having the same objects as $\underline{S}$, called the Poincaré category of $\underline{S}$.

For example, if $\underline{A}$ is an abelian category, and $\underline{\underline{Ext}}_{\underline{A}}$ is the bicategory of extensions in $\underline{A}$ (cf. (2.8)) then $\Pi\,\underline{Ext}_{\underline{A}}$, written $\mathrm{Ext}_{\underline{A}}$, is the category having $\underline{A}$ as set of objects, with maps the equivalence classes of extensions under the usual equivalence relation.

If $\Phi = (F, \varphi): \underline{S} \longrightarrow \overline{\underline{S}}$ is a morphism of bicategories, we define a functor $\Pi\Phi: \Pi\underline{S} \longrightarrow \Pi\overline{\underline{S}}$ by $\Pi\Phi A = FA$ and $\Pi\Phi[S]_\Pi = [FS]_\Pi$. Thus as a map of diagrams $\Pi\Phi$ depends only on the F part, but it is a functor thanks to the φ part which connects $FS \circ FT$ and $F(S \circ T)$ and also FI_A and I_{FA}. We clearly define thus the Poincaré functor

$$\Pi: \mathrm{Bicat}^{[1]} \longrightarrow \mathrm{Cat}^{[1]} .$$

(7.1.1) Proposition.

(i) The Poincaré functor is left adjoint to the degeneracy functor

$$D: \mathrm{Cat}^{[1]} \longrightarrow \mathrm{Bicat}^{[1]} \text{ of } (5.1).$$

(ii) The composite $\Pi \circ D$ is isomorphic to the identity functor of $\mathrm{Cat}^{[1]}$.

The proof, straightforward, is omitted.

<u>(7.2) Classifying category.</u> In many cases the equivalence relation defining $\Pi\underline{S}$ is too coarse. Thus if all the $\underline{S}(A,B)$ are connected and non-empty (e.g. for $\underline{S} = \text{Bim}$) the category $\Pi\underline{S}$ is equivalent to one point. A more precise category is defined as follows: Let $\underline{S}$ be a bi-category. For each object S of $\underline{S}(A,B)$ let $[S]$ be the set of all objects of $\underline{S}(A,B)$ isomorphic to S. Define $C\underline{S}(A,B)$ to be the set of all such isomorphism classes. We have a composition:

$$C\underline{S}(A,B) \times C\underline{S}(B,C) \longrightarrow C\underline{S}(A,C) \qquad ([S],[T]) \rightsquigarrow [S \circ T]$$

giving rise to a category $C\underline{S}$ having same objects as $\underline{S}$ called the <u>classifying category</u> of $\underline{S}$.

If $\underline{S} = \text{Cat}$, $C\underline{S}$ is the category with objects the categories, and maps isomorphism classes of functors; if $\underline{S}$ is the c.m. of modules over a commutative ring Λ, $C\underline{S}$ is the monoid with elements classes of isomorphic modules and composition induced by $\otimes$, etc....

The category $C\underline{S}$ is less functorial than the Poincare category: If $\Phi = (F,\varphi): \underline{S} \longrightarrow \overline{\underline{S}}$ is a morphism, the correspondence $A \rightsquigarrow FA$ $[S] \rightsquigarrow [FS]$ defines a map of the underlying graphs of $C\underline{S}$ and $C\overline{\underline{S}}$, however $[FS]\circ[FT] \neq [F(S\circ T)]$. However, if Φ is a homomorphism this map is a functor $C\Phi: C\underline{S} \longrightarrow C\overline{\underline{S}}$. Thus we obtain a classifying functor

$$C\colon \text{H-Bicat}^{[1]} \longrightarrow \text{Cat}^{[1]}.$$

Clearly, we have a natural surjection $C\underline{S} \longrightarrow \Pi\underline{S}$ which is an isomorphism when $\underline{S}$ is locally a groupoid (i.e., all the $\underline{S}(A,B)$'s are groupoids).

(7.3) Picard groupoid. If $\underline{S}$ is a bicategory, the invertible maps of the classifying category $C\underline{S}$ form a groupoid $\mathrm{Pic}\,\underline{S}$, called the Picard groupoid of $\underline{S}$. Clearly we obtain the Picard functor

$$\mathrm{Pic}: \text{H-Bicat}^{[1]} \longrightarrow \text{Groupoid}^{[1]}.$$

The definition is motivated by:

(7.3.1) Theorem: Let R be a commutative ring with identity, Mod(R) the c.m. of R-modules (with $\otimes_R$ as multiplication). Then Pic Mod(R) is canonically isomorphic to the Picard group of R, Pic R.

All there is to show is that, if M is an R-module such that there exists an R-module N with $M \otimes N \simeq R$ and $N \otimes M \simeq R$, then M is finitely generated projective. The proof is left to the reader since it will result from general considerations of Part II.

(7.4) Inverse limits. The general notion of limits of bicategories shall be examined in Part II, in connection with bi-adjoints. We will need immediately the following:

(7.4.1) Proposition: (i) The category $\text{S}\circ\text{H-Bicat}^{[1]}$ has inverse limits (and even a canonical choice of limits). (ii) The inclusion functors of $\text{S}\circ\text{H-Bicat}^{[1]}$ in $\text{H-Bicat}^{[1]}$, $\text{S}\circ\text{U-Bicat}^{[1]}$, $\text{U-Bicat}^{[1]}$ and $\text{Bicat}^{[1]}$ commute with the inverse limits.

Proof. (i) follows from the fact that bicategories are algebraic structures (cf. (1.4)(iii)) and that their morphisms as algebraic structures are the strict homomorphisms (cf. (4.2)). If $\underline{T}$ is an indexing category and $\underline{S}_i$ a family of bicategories indexed by $\underline{T}$, the transition maps being strict homomorphisms, $\varprojlim \underline{S}_i = \underline{S}$ is constructed pointwise, i.e., $\mathrm{Ob}\,\underline{S} = \varprojlim \mathrm{Ob}\,\underline{S}_i$; $\underline{S}(A,B) = \varprojlim \underline{S}_i(A_i, B_i)$ for $A = (A_i)$, $B = (B_i)$, etc..., the maps $\underline{S} \longrightarrow \underline{S}_i$ are the obvious projections. The proof of (ii) is straightforward and is omitted.

8. Transformations between Morphisms

(8.1) Introduction. Starting with categories, which are one-dimensional graphs with one operation, we get for the system of "all possible maps" (functors and natural transformations) a bicategory Cat which is a 2-dimensional complex with two operations. Similarly, "all the maps" between bicategories should constitute a 3-dimensional complex with three (partially defined) operations. Apart from internal coherence the examples given in §5 already oblige us to construct completely this 3-dimensional structure: We have shown that many notions usually thought of as objects -- e.g., algebras, categories, monads, ... -- could be identified with morphisms of bicategories $\Phi: \underline{S} \longrightarrow \underline{S}'$ for suitable $\underline{S}$ and $\underline{S}'$. However, if Φ and Ψ are two such objects, there are usually maps between them which sould correspond to transformations between morphisms of bicategories, i.e., 2-cells. Moreover, if Φ and Ψ were categories, the functors $\Phi \longrightarrow \Psi$ would give 2-cells, but we would, and will indeed, interpret natural transformations as 3-cells of Bicat.

To construct the 2-dimensional skeleton $\text{Bicat}^{[2]}$ of Bicat we use the following idea of category theory: If f_o and f_1 are functors $\underline{X} \longrightarrow \underline{Y}$ a natural transformation can be defined in either of these two ways (*) :

(i) A functor $h: \mathbb{2} \times \underline{X} \longrightarrow \underline{Y}$ such that $h \circ (\partial_i \times \text{Id}) = f_i$ $(i = 0, 1)$ where $\partial_i : \mathbb{1} \longrightarrow \mathbb{2}$ are the obvious functors.

(*) Compare with the definition of a homotopy by $I \times X \longrightarrow Y$ or $X \longrightarrow Y^I$.

(ii) A functor $k\colon \underline{X} \longrightarrow \underline{Y}^{\mathbf{2}}$ such that $d_i k = f_i$ where

$$d_i\colon \underline{Y}^{\mathbf{2}} \xrightarrow{Y^{\partial_i}} \underline{Y}^{\mathbf{1}} \longrightarrow \underline{Y}\,.$$

However none of these definitions suffices to define the composition of natural transformations $f_0 \longrightarrow f_1 \longrightarrow f_2$. It is obtained by means of

(i) A functor $\gamma\colon \mathbf{2} \longrightarrow \mathbf{2} \coprod_{(\partial_0, \partial_1)} \mathbf{2}$ or

(ii) A functor $c\colon \underline{Y}^{\mathbf{2}} \prod_{(d_0, d_1)} \underline{Y}^{\mathbf{2}} \longrightarrow \underline{Y}^{\mathbf{2}}$

such that $\mathbf{2}$ is a cocategory inside $\mathrm{Cat}^{[1]}$ (resp. $\underline{Y}^{\mathbf{2}}$ is a category inside $\mathrm{Cat}^{[1]}$). In $\mathrm{Cat}^{[1]}$ the passage from (i) to (ii) is trivial, but it is far from being so in $\mathrm{Bicat}^{[1]}$, and the analogue of (ii), being less complicated, will be used. Thus, the aim of the section is to assign to each bicategory $\underline{S}$ a bicategory called $\mathrm{Cyl}\underline{S}$, equipped with strict homomorphisms $d_0, d_1\colon \mathrm{Cyl}\underline{S} \longrightarrow \underline{S}$ and $c\colon \mathrm{Cyl}\underline{S} \prod_{(d_0, d_1)} \mathrm{Cyl}\underline{S} \longrightarrow \mathrm{Cyl}\underline{S}$ (the pullback exists because of (7.4.1)). $\mathrm{Cyl}\underline{S}$ plays the same universal role in this context as the space of paths in topology.

<u>(8.2) Squares and cylinders.</u> Let $\underline{S}$ be a bicategory, $U\colon A \longrightarrow \bar{A}$ and $V\colon B \longrightarrow \bar{B}$ be arrows of $\underline{S}$.

(8.2.1) A <u>square from V to U</u> $Q = (\bar{S}, u, S)\colon V \longrightarrow U$ is defined (*) by:

(i) two arrows $S\colon B \longrightarrow A$ and $\bar{S}\colon \bar{B} \longrightarrow \bar{A}$.

(ii) a 2-cell $u\colon \bar{S} \circ V \Longrightarrow U \circ S$ (i.e., an arrow of $\underline{S}(\bar{A}, B)$).

(*) See (8.5) below for a geometric interpretation.

The square Q is said commutative if $U \circ S = \overline{S} \circ V$ and u is the identity, commutative up to isomorphism, or iso-commutative if u is invertible (in $\underline{S}(\overline{A}, B)$).

Let $Q_1 = (\overline{S}_1, u_1, S_1): V \longrightarrow U$ and $Q_2 = (\overline{S}_2, u_2, S_2): V \longrightarrow U$ be two squares with the same domain V and codomain U.

(8.2.2) A cylinder from Q_2 to Q_1, $q = (\overline{s}, s): Q_2 \longrightarrow Q_1$ is defined (*) by a pair of 2-cells:

$$S_1 \xleftarrow{\ s\ } S_2 \text{ in } \underline{S}(A, B) \text{ and } \overline{S}_1 \xleftarrow{\ \overline{s}\ } \overline{S}_2 \text{ in } \underline{S}(\overline{A}, \overline{B})$$

making the following diagram of $\underline{S}(\overline{A}, B)$ commutative:

(8.2.3)
$$\begin{array}{ccc} \overline{S}_1 \circ V & \xleftarrow{\ \overline{s} \circ V\ } & \overline{S}_2 \circ V \\ {\scriptstyle u_1}\downarrow & & \downarrow{\scriptstyle u_2} \\ U \circ S_1 & \xleftarrow[\ U \circ s\ ]{} & U \circ S_2 \end{array}$$

that is, satisfying the equation:

(Cyl): $$(U \circ s)u_2 = u_1(\overline{s} \circ V) .$$

(8.3) The categories $\mathrm{Cyl}\,\underline{S}(U, V)$: Let $Q^i = (\overline{S}^i, u^i, S^i): V \longrightarrow U$ $(i = 1, 2, 3)$ be three squares with same domain and codomain, and $q^j = (\overline{s}^j, s^j): Q^{j+1} \longrightarrow Q^j$ $(j = 1, 2)$ be two cylinders, then the composite $\overline{s}^1 \overline{s}^2$ and $s^1 s^2$ are defined, and we have:

(*) See (8.5) below for a geometric interpretation.

(8.3.1) Lemma. (i) The pair $(\bar{s}^1\bar{s}^2, s^1s^2)$ defines a cylinder from Q^3 to Q^1 written q^1q^2.

(ii) With the composition $(q^1, q^2) \leadsto q^1q^2$ we obtain a category, denoted $\mathrm{Cyl}\,\underline{S}(U, V)$, having as objects the squares from V to U, and as maps the cylinders between these squares. If $Q = (\bar{S}, u, S): V \longrightarrow U$ is an object of $\mathrm{Cyl}\,\underline{S}(U, V)$, its identity is the cylinder i_Q defined by $(i_{\bar{S}}, i_S)$.

Proof. The equation (Cyl) for $(\bar{s}^1\bar{s}^2, s^1s^2)$ is equivalent to the commutativity of the outside of the following diagram in $\underline{S}(\bar{A}, B)$

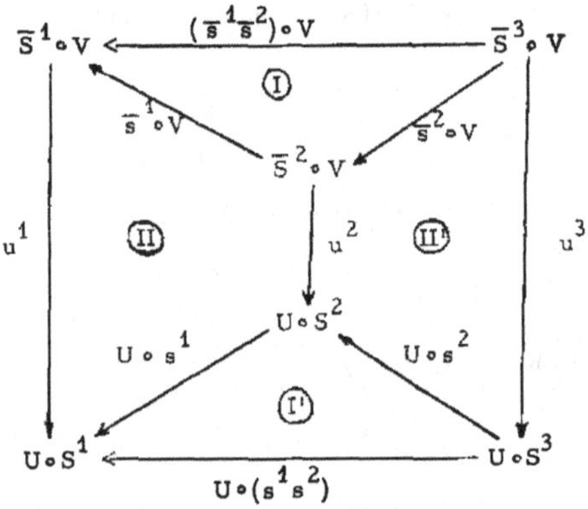

But the regions numbered I and I' commute because $\circ$ are bifunctors, and II and II' because q^1 and q^2 are cylinders. This proves (i), then (ii) follows trivially from the fact that $\underline{S}(A, B)$ and $\underline{S}(\bar{A}, \bar{B})$ are categories.

(8.4) The functors $c(U, V, W)$. Let $U: A \longrightarrow \bar{A}$, $V: B \longrightarrow \bar{B}$ and $W: C \longrightarrow \bar{C}$ be arrows of a bicategory $\underline{S}$.

(8.4.1). If $Q = (\overline{S}, u, S): V \longrightarrow U$ and $R = (\overline{T}, v, T): W \longrightarrow V$ are two squares, we define a square $Q \circ R$ from W to U by

$$Q \circ R = (\overline{S} \circ \overline{T}, u/v, S \circ T): W \longrightarrow U$$

where u/v is the composite map in $\underline{S}(\overline{A}, C)$:

$$U \circ (S \circ T) \xleftarrow{a} (U \circ S) \circ T \xleftarrow{u \circ T} (\overline{S} \circ V) \circ T \xleftarrow{a^{-1}} \overline{S} \circ (V \circ T) \xleftarrow{S \circ v} \overline{S} \circ (\overline{T} \circ W) \xleftarrow{a} (\overline{S} \circ T) \circ W.$$

(8.4.2). Suppose we are given furthermore two squares:

$$Q_1 = (\overline{S}_1, u_1, S_1): V \longrightarrow U \quad \text{and} \quad R_1 = (\overline{T}_1, v_1, T_1): W \longrightarrow V$$

and two cylinders:

$$q = (\overline{s}, s): Q_1 \longrightarrow Q \quad \text{and} \quad r = (\overline{t}, t): R_1 \longrightarrow R$$

then the composites $s \circ t$ and $\overline{s} \circ \overline{t}$ are defined and we have:

<u>(8.4.3) Lemma:</u> (i) The pair $(\overline{s} \circ \overline{t}, s \circ t)$ determines a cylinder from $Q_1 \circ R_1$ to $Q \circ R$, written $q \circ r$.

(ii) The composition $(Q, R) \rightsquigarrow Q \circ R$, $(q, r) \rightsquigarrow q \circ r$ is a bifunctor:

$$c(U,V,W): \mathrm{Cyl}\,\underline{S}(U, V) \times \mathrm{Cyl}\,\underline{S}(V, W) \longrightarrow \mathrm{Cyl}\,\underline{S}(U, W).$$

<u>Proof.</u> To show (i) we must prove that the equation (Cyl) holds for $(\overline{s} \circ \overline{t}, s \circ t)$ and the squares $Q_1 \circ R_1$ and $Q \circ R$, which means, according to the definition of $Q \circ R$, that the exterior of the following diagram commutes:

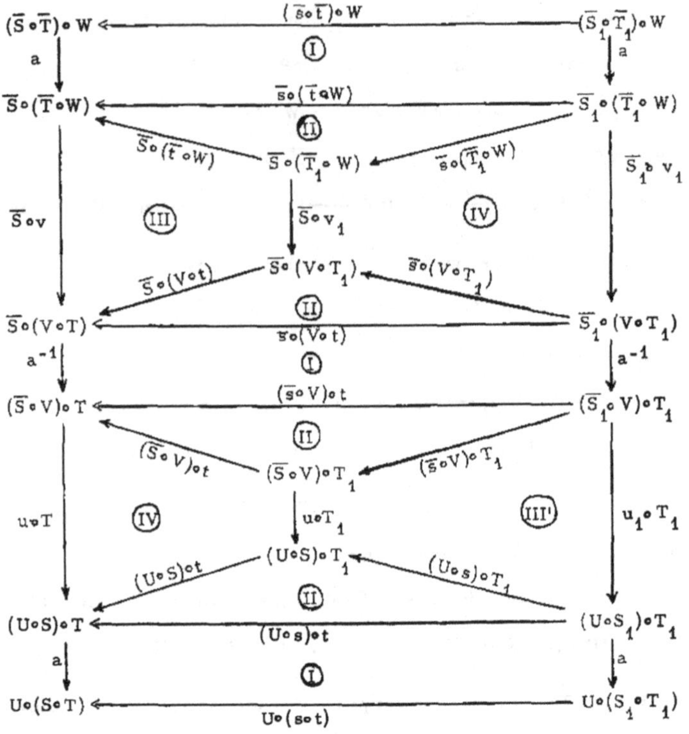

<u>(8.5) Geometric representation.</u> The definitions and results of (8.1) to (8.4) admit the following geometric interpretation

(8.5.1) A square $Q = (\overline{S}, u, S)\colon V \longrightarrow U$ can be represented by:

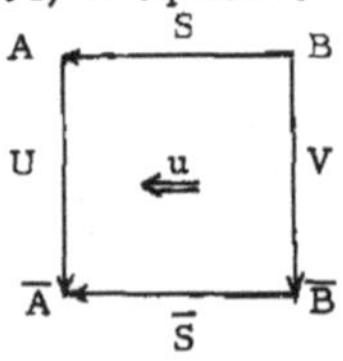

abbreviated in

(8.5.2) A cylinder $q: (\bar{s}, s): Q_2 \Longrightarrow Q_1$, by :

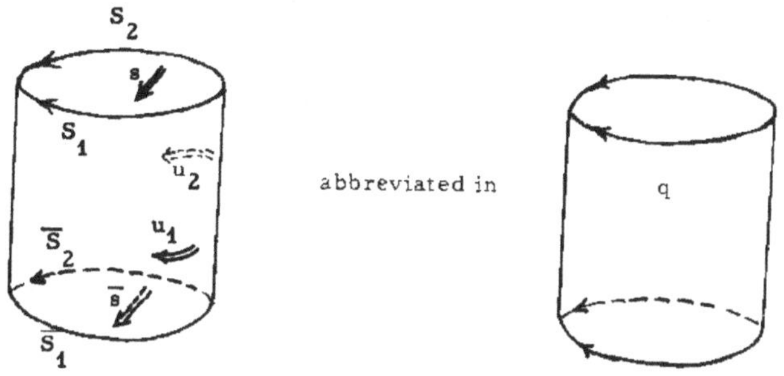

(8.5.3) The part (i) of the Lemma (8.3.1) means that cylinders can be pasted according to the picture:

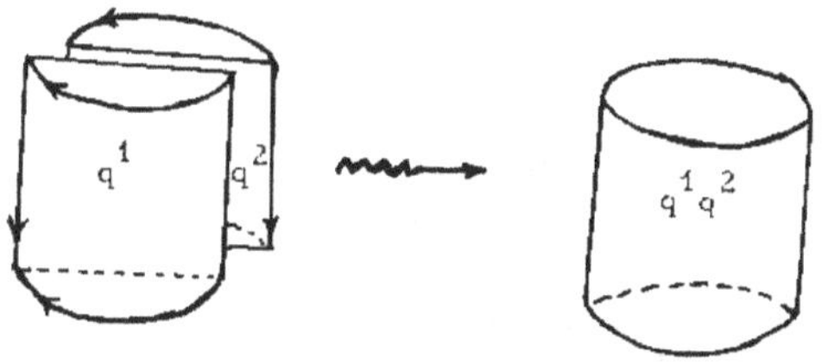

and the part (ii) essentially means that this pasting is associative.

(8.5.4) The composition $(Q, R) \rightsquigarrow Q \circ R$ of (8.4.1) corresponds to:

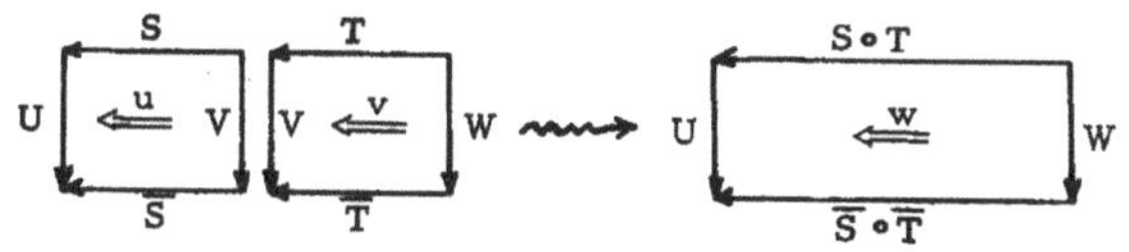

(8.5.5) The part (i) of Lemma (8.4.3) means that cylinders q and r can be pasted along V to get a new cylinder:

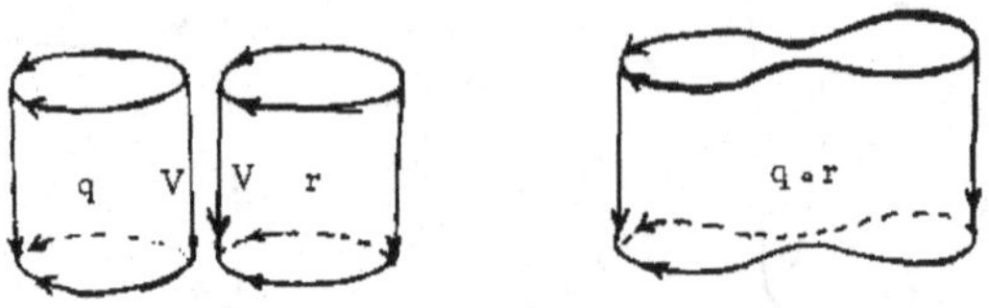

And part (ii) means essentially that the following "diagram" is commutative

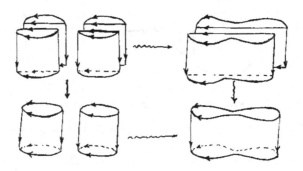

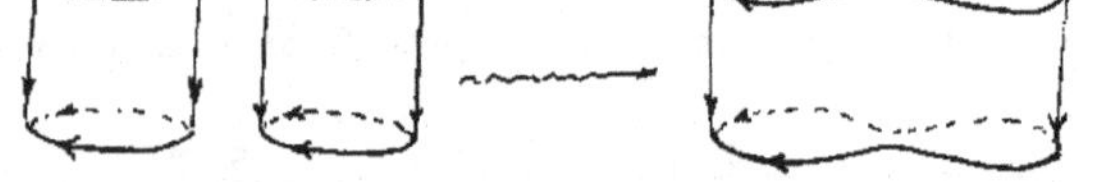

In what follows we shall use frequently this geometrical representation which motivates and makes comprehensible definitions such as (8. 4. 1) or (8.4.2), and makes plausible results such as (8. 3. 1) or (8.4.3). However, the suggestions of geometry cannot replace proofs, and should be taken with a "grain of salt" because of the lack of associativity of $\circ$; thus the pairing $(q, r) \longrightarrow q \circ r$ of cylinders is not associative, and neither is their superposition (8. 8). Nevertheless, to avoid diagrams such as (8. 6. 2) we shall replace many proofs by their geometrical analogs.

<u>(8. 6) The associativity and identity isomorphisms.</u> Let $\underline{S}$ be a bicategory and $Q_i = (\overline{S}_i, u_i, S_i): U_i \longrightarrow U_{i-1}$ be three squares satisfying the incidence relations depicted by:

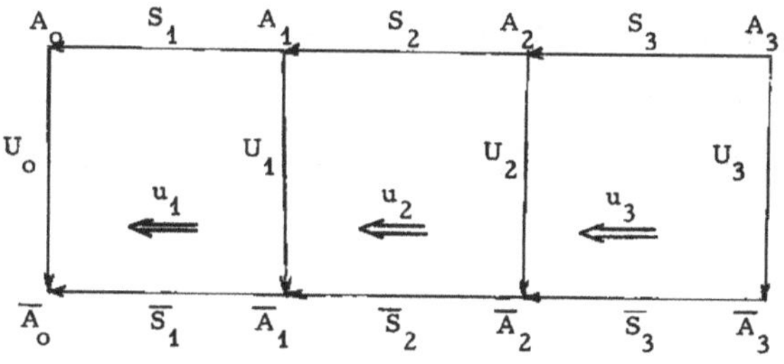

They determine:

(i) Two, in general distinct, squares from U_3 to U_o :

$$(Q_1 \circ Q_2) \circ Q_3 \qquad \text{and} \qquad Q_1 \circ (Q_2 \circ Q_3)$$

(ii) Two isomorphisms, in the categories $\underline{S}(A_o, A_3)$ and $\underline{S}(\overline{A}_o, \overline{A}_3)$:

$$a(A_o, A_1, A_2, A_3)(S_1, S_2, S_3): (S_1 \circ S_2) \circ S_3 \xrightarrow{\sim} S_1 \circ (S_2 \circ S_3)$$

$$a(\overline{A}_o, \overline{A}_1, \overline{A}_2, \overline{A}_3)(\overline{S}_1, \overline{S}_2, \overline{S}_3): (\overline{S}_1 \circ \overline{S}_2) \circ \overline{S}_3 \xrightarrow{\sim} \overline{S}_1 \circ (\overline{S}_2 \circ \overline{S}_3)$$

satisfying the incidence relations depicted by:

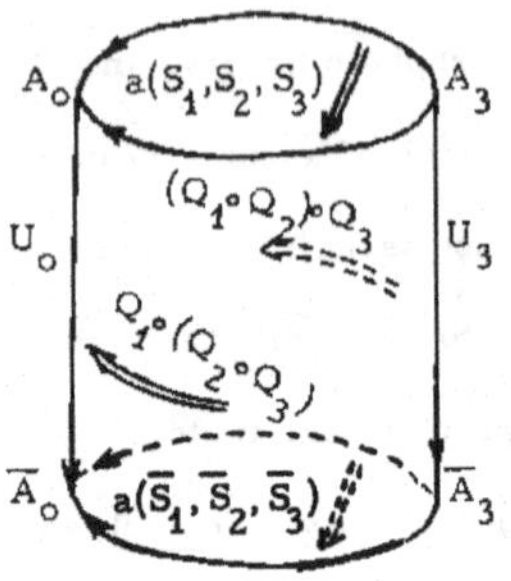

(8.6.1) Lemma. With the previous notations:

(i) The pair $(a(A_o, A_1, A_2, A_3)(S_1, S_2, S_3)\ ,\ a(\overline{A}_o, \overline{A}_1, \overline{A}_2, \overline{A}_3)(\overline{S}_1, \overline{S}_2, \overline{S}_3))$ defines a cylinder:

$$a(U_o, U_1, U_2, U_3)(Q_1, Q_2, Q_3): (Q_1\circ Q_2)\circ Q_3 \longrightarrow Q_1\circ(Q_2\circ Q_3).$$

(ii) For U_i fixed, the family $a(U_o, U_1, U_2, U_3)(Q_1, Q_2, Q_3)$ is functorial in the Q_j's, that is, determines a natural transformation:

$$a(U_o, U_1, U_2, U_3): c(U_o, U_2, U_3)\,(c(U_o, U_1, U_2)\times \mathrm{Id}) \to c(U_o, U_1, U_3)\,(\mathrm{Id}\times c(U_1, U_2, U_3))$$

between the composite functors bounding the diagram:

$$\begin{array}{ccc}
\mathrm{Cyl}\,\underline{S}(U_o, U_1)\times \mathrm{Cyl}\,\underline{S}(U_1, U_2)\times \mathrm{Cyl}\,\underline{S}(U_2, U_3) & \xrightarrow{\mathrm{Id}\times c(U_1, U_2, U_3)} & \mathrm{Cyl}\,\underline{S}(U_o, U_1)\times \mathrm{Cyl}\,\underline{S}(U_1, U_3) \\
\Big\downarrow{\scriptstyle c(U_o, U_1, U_2)\times \mathrm{Id}} & & \Big\downarrow{\scriptstyle c(U_o, U_1, U_3)} \\
\mathrm{Cyl}\,\underline{S}(U_o, U_2)\times \mathrm{Cyl}\,\underline{S}(U_2, U_3) & \xrightarrow[c(U_o, U_2, U_3)]{} & \mathrm{Cyl}\,\underline{S}(U_o, U_3)
\end{array}$$

(iii) The natural transformation $a(U_o, U_1, U_2, U_3)$ are isomorphisms.

(iv) The $a(U_o, U_1, U_2, U_3)$, for the U_i's ranging in the arrows of $\underline{S}$, satisfy the associativity coherence (A.C).

Proof. Remembering the definition of u/v and $Q \circ R$ in (8.4.1) the equation (Cyl) is here equivalent to the commutativity of the exterior of diagram (8.6.2) below.

But the regions called I commute because the natural transformations a's of $\underline{S}$ satisfy (A.C) of §1; the II's by naturality of the a's ; the III's by definition of u/v; IV is the image of the diagram defining u_2/u_3 by the functor $S_1 \circ (\ \ )$ and similarly IV' "is" $(u_1/u_2) \circ S_3$. This proves (i). Then (ii), (iv) follow immediately from the similar statements which hold in $\underline{S}$ for $a(A_o, A_1, A_2, A_3)$ and $a(\overline{A}_o, \overline{A}_1, \overline{A}_2, \overline{A}_3)$. Finally (iii) will hold if we know that the pair:

$([a(A_o, A_1, A_2, A_3)(S_1, S_2, S_3)]^{-1}, [a(\overline{A}_o, \overline{A}_1, \overline{A}_2, \overline{A}_3)(\overline{S}_1, \overline{S}_2, \overline{S}_3)]^{-1})$ defines a cylinder from $Q_1 \circ (Q_2 \circ Q_3)$ to $(Q_1 \circ Q_2) \circ Q_3$, but this follows from (i) by conjugation.

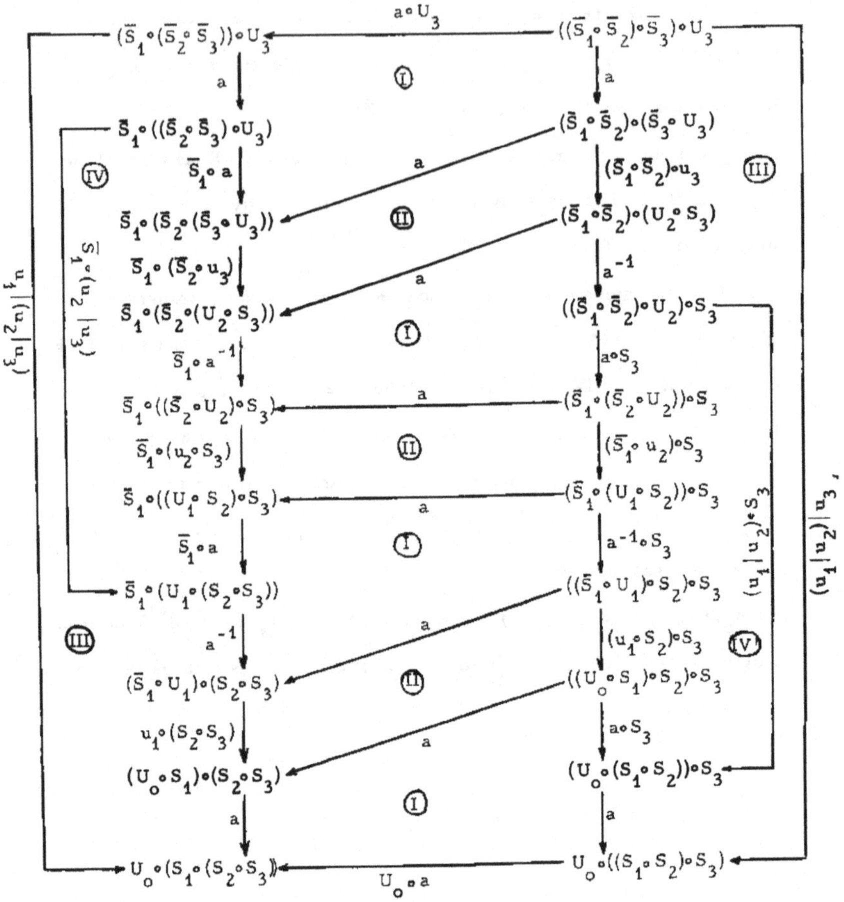

Diagram (8. 6. 2)

(8.6.3). Let $U: A \longrightarrow \bar{A}$ be an arrow of $\underline{S}$. We define a square I_U from U to U by $I_U = (I_{\bar{A}}, k_U, I_A): U \longrightarrow U$, where k_U is the unique arrow of the category $\underline{S}(\bar{A}, A)$ making commutative the diagram

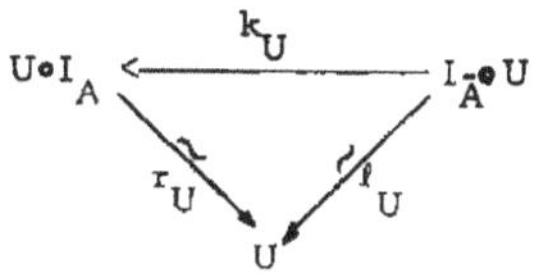

(8.6.4). Let $Q = (\bar{S}, u, S): V \longrightarrow U$ be a square. According to the picture

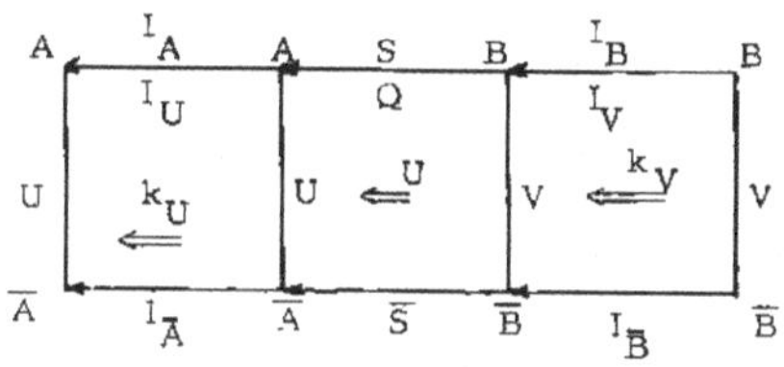

the square Q determines:

(i) Two squares $I_U \circ Q$ and $Q \circ I_V$ from V to U.

(ii) Four isomorphisms

$\ell(A,B)(S): I_A \circ S \xrightarrow{\sim} S$ and $r(A,B)(S): S \circ I_B \xrightarrow{\sim} S$ in $\underline{S}(A,B)$

$\ell(\bar{A},\bar{B})(\bar{S}): I_{\bar{A}} \circ \bar{S} \xrightarrow{\sim} \bar{S}$ and $r(\bar{A},\bar{B})(\bar{S}): \bar{S} \circ I_{\bar{B}} \xrightarrow{\sim} \bar{S}$ in $\underline{S}(\bar{A},\bar{B})$.

<u>(8.6.5) Lemma.</u> With the previous notations:

(i) The pairs $(\ell(A,B)(S), \ell(\bar{A},\bar{B})(\bar{S}))$ and $(r(A,B)(S), r(\bar{A},\bar{B})(\bar{S}))$ define two cylinders:

$\ell(U,V)(Q): I_U \circ Q \longrightarrow Q$ and $r(U,V)(Q): Q \circ I_V \longrightarrow Q$.

(ii) For U and V fixed, the families $\ell(U,V)(Q)$ and $r(U,V)(Q)$ are functorial in Q, that is, determine natural transformations $\ell(U,V)$ and $r(U,V)$ as shown in the diagram:

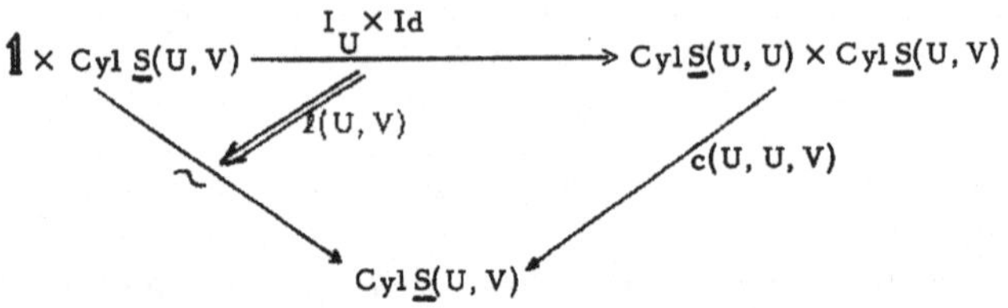

and the obvious analogue for $r(U,V)$.

(iii) The $r(U,V)$ and $\ell(U,V)$ are isomorphisms.

(iv) The system of natural transformations $a(U_0, U_1, U_2, U_3)$, $\ell(U,V)$ and $r(U,V)$ satisfy the coherence axiom (I. C) of 1.

The proof, completely similar to that of (8. 6. 1) except for smaller diagrams, is left to the reader.

(8. 6. 6). If $\underline{S}$ is a bicategory we define $\mathrm{Ob}(\mathrm{Cyl}\,\underline{S})$ to be the set of arrows of $\underline{S}$.

Putting together the Lemmas (8. 3. 1), (8. 4. 3), (8. 6. 1), and (8. 6. 5), we obtain:

<u>(8. 6. 7) Theorem:</u> The data: $\mathrm{Ob}(\mathrm{Cyl}\,\underline{S})$, $\mathrm{Cyl}\,\underline{S}(U,V)$, $c(U,V,W)$, I_U, $a(U_0, U_1, U_2, U_3)$, $\ell(U,V)$, $r(U,V)$ determine a bicategory $\mathrm{Cyl}\,\underline{S}$ called <u>bicategory of cylinders</u> of $\underline{S}$

(8.6.8) Remark: In all the steps of the passage from $\underline{S}$ to Cyl $\underline{S}$ there appears a shift of dimension in the definitions as well as the coherence properties involved: Thus to define the binary operations $(Q, R) \rightsquigarrow Q \circ R$ in Cyl $\underline{S}$, we need the existence in $\underline{S}$ of the ternary associativity isomorphisms $a(S, T, U)$. To define a in Cyl $\underline{S}$, we need to know that a is coherent up to the order 4 (see proof of (8.6.1)(i)). This suggests the conditions to require in higher dimensional cases (*).

(8.7) The top and bottom homomorphisms Let $\underline{S}$ be a bicategory, we define

(i) Two maps ∂_i: Ob Cyl $\underline{S} \longrightarrow$ Ob $\underline{S}$, $i = 0, 1$, by

$$\partial_0 U = A, \ \partial_1 U = \overline{A} \quad \text{for each arrow } U: A \longrightarrow \overline{A} \text{ of } \underline{S}$$

(ii) Two families of functors, indexed by pairs (U, V) of arrows of $\underline{S}$

$$\partial_i(U, V): \text{Cyl}\,\underline{S}(U, V) \longrightarrow \underline{S}(\partial_i U, \partial_i V), \quad Q \rightsquigarrow \partial_i Q, \quad q \rightsquigarrow \partial_i q \quad \text{by:}$$

$$\partial_0 Q = S, \ \partial_1 Q = \overline{S}, \quad \partial_0 q = s, \ \partial_1 q = \overline{s} \quad \text{for } Q \text{ and } q \text{ as below:}$$

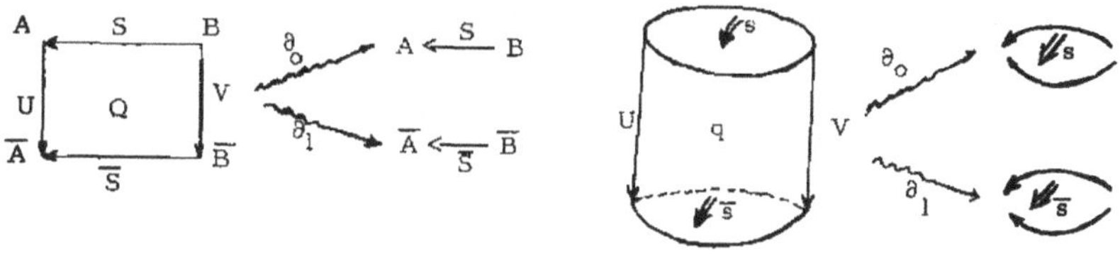

It follows from the definitions that the ∂_i commute strictly with the $\circ$'s and I's and define thus strict homomorphisms called respectively top and bottom

$$\partial_0 \text{ and } \partial_1 : \text{Cyl}\,\underline{S} \longrightarrow \underline{S}.$$

(*) Compare with $\Pi_i(\Omega X) \approx \Pi_{i+1}(X)$ in homotopy theory (see footnote p. 57).

<u>(8. 8) Superposition morphism.</u> Let $Q = (\bar{S}, u, S): V \longrightarrow U$ and $\bar{Q} = (\bar{\bar{S}}, \bar{u}, \bar{S}): \bar{V} \longrightarrow \bar{U}$ be two squares such that $\partial_o \bar{Q} = \bar{S} = \partial_1 Q$, (see picture (8.8.1) below). In the category $\underline{S}(\bar{\bar{A}}, B)$ we have the composite map, written $\bar{u} * u$:

$$(\bar{u}\circ u)\circ S \xleftarrow{a^{-1}} \bar{U}\circ(U\circ S) \xleftarrow{\bar{U}\circ u} \bar{U}\circ(\bar{S}\circ V) \xleftarrow{a} (\bar{U}\circ\bar{S})\circ V \xleftarrow{\bar{u}\circ V} (\bar{\bar{S}}\circ\bar{V})\circ V \xleftarrow{a^{-1}} \bar{\bar{S}}\circ(\bar{V}\circ V)$$

which determines a square $\bar{Q} * Q = (\bar{\bar{S}}, \bar{u} * u, S): \bar{V}\circ V \longrightarrow \bar{U}\circ U$, called superposition of Q on $\bar{Q}$ (along $\bar{S}$).

(8.8.1) 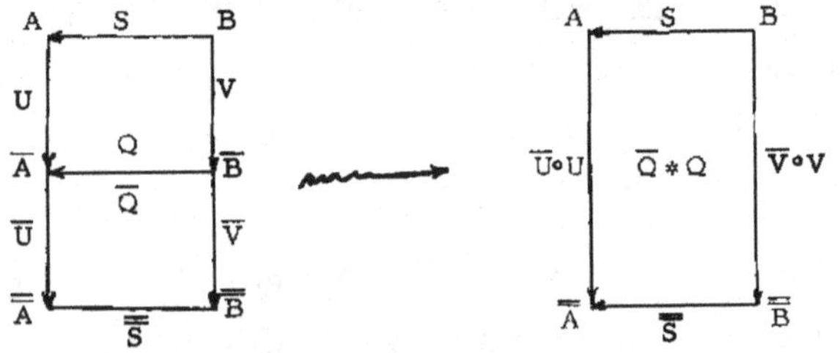

Similarly, let $q = (\bar{s}, s): Q_1 \longrightarrow Q$ and $\bar{q} = (\bar{\bar{s}}, s): \bar{Q}_1 \longrightarrow \bar{Q}$ be two cylinders such that $\partial_o \bar{q} = \bar{s} = \partial_1 q$, as depicted below:

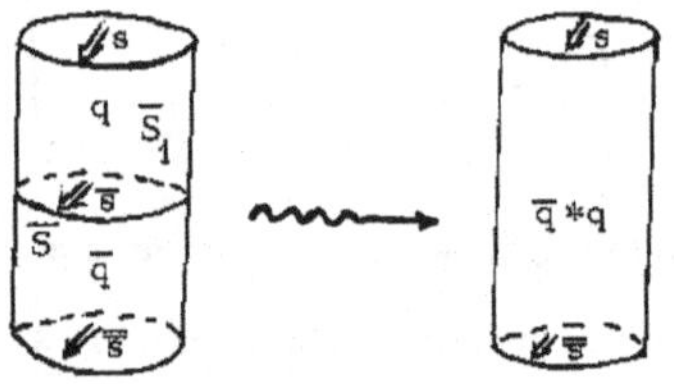

Using a diagram similar to (8. 6. 2) one can prove:

<u>(8. 8. 2) Lemma.</u> The pair $(\bar{\bar{s}}, s)$ defines a cylinder from $\bar{Q}_1 * Q_1$ to $\bar{Q} * Q$ written $\bar{q}\,{}_{\bar{s}}\,q$ and called superposition of q on $\bar{q}$ (along $\bar{s}$).

(8.8.3) Proposition. The superposition of cylinders and squares define a strict homomorphism

$$* : \text{Cyl}\,\underline{S} \prod_{(\partial_o,\partial_1)} \text{Cyl}\,\underline{S} \longrightarrow \text{Cyl}\,\underline{S}.$$

We have to show that the equations (i) and (ii) below, which state that * commutes with the compositions, hold (and similar equations for the identities, left to the reader.)

(i) $(\bar{q}'*q')(\bar{q}*q) = (\bar{q}'\bar{q})*(q'q)$

(ii) $(\bar{q}*q)\circ(\bar{r}*r) = (\bar{q}\circ\bar{r})*(q\circ r)$

whenever both sides are defined.

The equation (i) corresponds to the commutativity of:

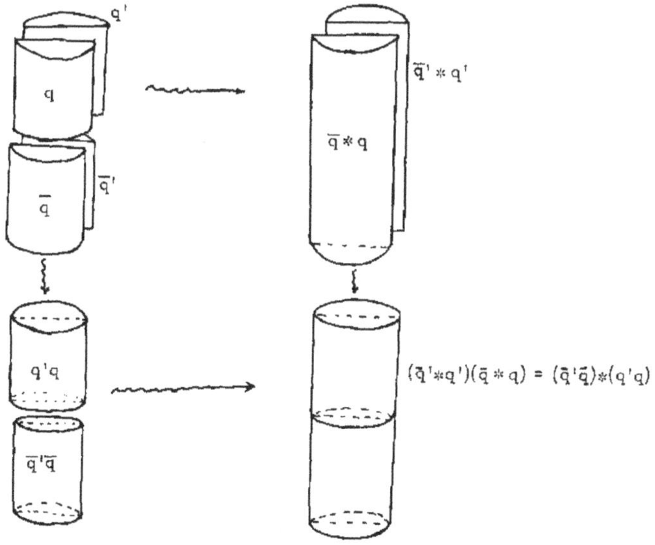

Similarly (ii) means that the pasting of the four cylinders

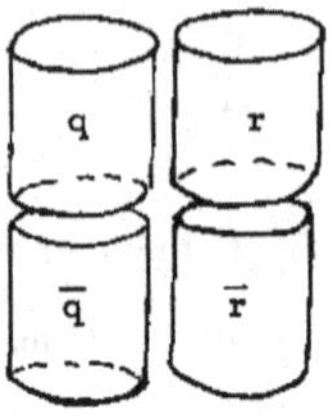

does not depend on the order. Proofs are omitted.

(B.1) Bénabou, J., Catégories avec multiplication, C. R. Acad. Sci., Paris 256 (1963), 1887-1890.

(B.2) ________, Algèbre élémentaire dans les catégories avec multiplication, C. R. Acad. Sci., Paris 258 (1964), 771-774.

(B.3) ________, Catégories relatives, C. R. Acad. Sci., Paris 260 (1965), 3824-3827.

(B.4) ________, Structures algébriques dans les catégories, to appear.

(E) Epstein, D., Steenrod operations in homological algebra, Invent. Math. 1 (1966), 152-208.

(EL) Ehresmann, C., Catégories structurées, Ann. Sci. Ecole Norm. Sup. 80 (1963), 349-425.

(EH) Eckmann-Hilton, Group-like structures in categories, Math. Ann. 145 (1962), 227-255.

(E. K) Eilenberg-Kelly, Closed categories, Proceedings of the La Jolla Conference on Categories, Springer 1967, 424-562.

(E. M) Eilenberg-Moore, Adjoint functors and triples, Illinois Journal of Mathematics, 9 (1965), 381-398.

(G) Godement, R., Théorie des faisceaux, Hermann Paris 1957.

(Gr) Grothendieck, A., Catégories fibrées et descente, Séminaire de Géometrie Algébrique 1961. Exposé VI.

(M) MacLane, S., Natural associativity and commutativity, Rice Univ. Studies 49 (1963), 28-46.

(Y) Yoneda, N., On Ext and exact sequences, J. Fac. Sci. Univ. Tokyo, Sec. I, 7 (1954), 193-227.

PROJECTIVE CLASSES AND ACYCLIC MODELS

A. Dold, S. MacLane[1], and U. Oberst

This brief note is to observe that the standard method of acyclic models ([5]) and the usual comparison theorems of absolute and relative homological algebra ([6],[9]) can all be subsumed under a single comparison theorem formulated for a projective class ([6],[11]) in an abelian category.

Let $\underline{A}$ be an abelian category. An object P of $\underline{A}$ is said to be projective relative to a morphism $e: B \longrightarrow C$ of $\underline{A}$ if to every $h: P \longrightarrow C$ there exists $h': P \longrightarrow B$ with $eh' = h$. As usual, this is equivalent to stating that

$$\underline{A}(P, e): \underline{A}(P, B) \longrightarrow \underline{A}(P, C)$$

is surjective, where $\underline{A}(P, B)$ stands for the set of morphisms from P to B. If E is any class of morphisms of $\underline{A}$, an object P of $\underline{A}$ is called $\underline{E}$-projective if and only if it is projective for all morphisms e of $\underline{E}$; in particular if $\underline{E}$ is the class of all epimorphisms of $\underline{A}$ then the $\underline{E}$-projectives are the usual projectives. For any $\underline{E}$, the usual proof shows that direct summands and direct sums of $\underline{E}$-projectives are $\underline{E}$-projective.

---

1. The research of the first and second named authors was supported in part by a grant from the National Science Foundation and the Office of Naval Research, respectively.

A complex K over an object A of $\underline{A}$ is, as usual, a sequence

$$\cdots \longrightarrow K_2 \xrightarrow{d_2} K_1 \xrightarrow{d_1} K_0 \xrightarrow{\varepsilon = d_0} A = K_{-1}$$

in which the composite of any two successive morphisms is zero. Let

$$i_n : Z_n \longrightarrow K_n \quad \text{for } n \geq 0$$

be the kernel of d_n. Since $d_n d_{n+1} = 0$, d_{n+1} factors as $d_{n+1} = i_n e_{n+1}$ with a unique $e_{n+1}: K_{n+1} \longrightarrow Z_n$. Call $K \xrightarrow{\varepsilon} A$ $\underline{E}$-acyclic if each e_n, $n \geq 0$, is in $\underline{E}$. If $\underline{E}$ is the class of epimorphisms of $\underline{A}$, this acyclicity is the usual acyclicity $\varepsilon : H_*(K) \cong A$. Call K $\underline{E}$-projective when every K_n, $n \geq 0$, is $\underline{E}$-projective. The standard argument (e.g. [9]) now proves

<u>Theorem 1:</u> (Comparison Theorem). Let $\underline{E}$ be any class of morphisms in an abelian category $\underline{A}$. Let $\varepsilon : K \longrightarrow A$ be an $\underline{E}$-projective and $\varepsilon' : K' \longrightarrow A'$ an $\underline{E}$-acyclic complex over A, resp. A'. Then to each morphism $f: A \longrightarrow A'$ there is a chain transformation $f_* : K \longrightarrow K'$ with $f\varepsilon = \varepsilon' f_o$, and any two such chain transformations are chain homotopic.

"Higher homotopies" may also be constructed as usual (e.g. A. DOLD [4]). Thus if

$$f_*, g_* : K \longrightarrow K'$$

are two such chain transformations and s_*, t_* two chain homotopies of f_* to g_*, there is a family of morphisms

$$h_n : K_n \longrightarrow K'_{n+2} \quad \text{for } n = 0, 1, 2, \ldots$$

with

$$d'_{n+2}h_n - h_{n-1}d_n = s_n - t_n \quad \text{for all } n.$$

More generally, the usual complex $\underline{A}(K, K')$ has zero homology in positive dimensions.

This comparison theorem contains the usual comparison theorem of homological algebra and the corresponding theorem for suitable classes of "proper" short exact sequences (e.g. [7], Th. III 6.1. and [9] IX.4.3). In many cases the class E has more special properties. Let $C(\underline{E})$, the closure of $\underline{E}$, consist of all those morphisms e' of $\underline{A}$ such that every $\underline{E}$-projective is also projective relative to e'. Thus $C(E) \supseteq E$. The class $\underline{E}$ is said to be closed if $C(\underline{E}) = \underline{E}$. We say A has enough $\underline{E}$-projectives if for every object A in $\underline{A}$ there is a morphism $e: P \longrightarrow A$ in $\underline{E}$ with $\underline{E}$-projective domain P. The class $\underline{E}$ is called a projective class if it is closed and if there are enough $\underline{E}$-projectives in $\underline{A}$. These notions are dual to those of S. EILENBERG, J.C. MOORE [6], and are compactly stated, with their relations to adjoint functors, in MITCHELL [12]. In J. MARANDAS terminology [11] the pair consisting of the projective class $\underline{E}$ and the class of all $\underline{E}$-projectives is called a projective structure.

We remark that a class $\underline{E}$ such that there are enough $\underline{E}$-projectives in $\underline{A}$ is a projective class if $\underline{E}$ satisfies the condition (PC): If a and e are two morphisms in A whose composite ea exists and lies in $\underline{E}$ then also e is an element of $\underline{E}$.

The method of acyclic models conventionally starts with a category $\underline{C}$ and a set $\underline{M}$ of objects of $\underline{C}$, called the "models"; its comparison theorem deals with natural transformations between functors on $\underline{C}$ to the category of complexes of abelian groups. We generalize the theory by replacing the category of abelian groups by an abelian category $\underline{A}$ with infinite direct sums and enough projectives, and work then in the abelian category $\underline{A}^{\underline{C}}$ of all functors $F: \underline{C} \longrightarrow \underline{A}$ (size is no real consideration; if you like, assume $\underline{C}$ small). The comparison by acyclic models follows from our previous comparison theorem by the following result.

Theorem 2: Let $\underline{M}$ be a set of objects in a category $\underline{C}$, and let $\underline{A}$ be an abelian category with infinite direct sums and enough projectives. In the category $\underline{A}^{\underline{C}}$ let $\underline{E}$ be the class of all those $e: F \longrightarrow G$ such that $e(M): F(M) \longrightarrow G(M)$ is an epimorphism in $\underline{A}$ for every model $M \in \underline{M}$. Then a complex $K \longrightarrow A$ over A is $\underline{E}$-acyclic if and only if, for every model M, $K(M) \longrightarrow A(M)$ is an acyclic complex in $\underline{A}$, while a functor F is $\underline{E}$- projective if and only if it is "representable" in the sense of acyclic models. Moreover $\underline{E}$ is a projective class.

Proof. The description of $\underline{E}$-acyclic complexes is immediate, and agrees with the usual requirement that the complex of functors be "acyclic on models". Moreover, $\underline{E}$ obviously satisfies the condition (PC) since the class of epimorphisms of $\underline{A}$ satisfies this condition.

Next we construct some $\underline{E}$-projectives. Write $\coprod$ for the coproduct of abelian groups. For any set X let

$$\underline{Z}X = \coprod \{\underline{Z}x; \; x \in X\}$$

be the free abelian group generated by X. For every model $M \in \underline{M}$ we obtain the functor

$$\underline{Z}\underline{C}(M, -): C \longmapsto \underline{Z}\,\underline{C}(M, C) = \coprod \{Z\gamma; \; \gamma \in \underline{C}(M, C)\}$$

in $\underline{Ab}^{\underline{C}}$. If, moreover, A is an object in $\underline{A}$, there is also the functor

(1) $$\underline{Z}\underline{C}(M, -) \otimes A: C \longmapsto \underline{Z}\,\underline{C}(M, C) \otimes A$$

in $\underline{A}^{\underline{C}}$, where, for any abelian group G and any object $\underline{A} \in A$, $G \otimes A$ denotes their tensor product. Here the tensor product

$$\otimes: \; \underline{Ab} \times \underline{A} \longrightarrow \underline{A}$$

is the unique biadditive functor which is right exact, commutes with direct sums, and is normalized by $\underline{Z} \otimes A = A$ for every $A \in \underline{A}$. For every $F \in \underline{A}^{\underline{C}}$ there is the isomorphism

(2) $$\underline{A}^{\underline{C}}(\underline{Z}\underline{C}(M, -) \otimes A, F) \longrightarrow \underline{A}(A, F(M)),$$

natural in $M \in \underline{M}$, $A \in \underline{A}$, and $F \in \underline{A}^{\underline{C}}$. Indeed, by the Yoneda lemma, the natural transformations indicated on the left of (2) are determined when the functors in question are evaluated at the object M and then $\underline{Z}\underline{C}(M, M)$ is replaced by the summand generated by the identity $1: M \longrightarrow M$. This gives the right side of (2). This proof is an application of the Yoneda lemma, so rests ultimately on the fact that $\underline{C}(M, -)$ is representable in the sense of GROTHENDIECK (see e. g. [10]). Now we claim that for projective P in $\underline{A}$ the functor $ZC(M,-)\otimes P$ is $\underline{E}$-projective. Indeed, let $e: F \longrightarrow G$ be in $\underline{E}$. Then $e(M): F(M) \longrightarrow G(M)$ is an epimorphism and thus

$$\underline{A}(P, e(M)): \underline{A}(P, F(M)) \longrightarrow \underline{A}(P, G(M))$$

is surjective since P is projective. By means of the isomorphism (2) this implies that

$$\underline{A}^{\underline{C}}(\underline{Z}\underline{C}(M,-)\otimes P, e)$$

is surjective, and hence $\underline{Z}\,\underline{C}(M,-)\otimes P$ is $\underline{E}$-projective.

Next let $P = (P(M))_{M \in \underline{M}}$ be any family of projective objects $P(M)$ in $\underline{A}$. Then the direct sum

$$\tilde{P} = \coprod \{\underline{Z}\,\underline{C}(M,-)\otimes P(M)\,;\ M \in \underline{M}\}$$

is also $\underline{E}$-projective. If $\underline{A}$ is the category of abelian groups then every $P(M)$ is free and hence has a basis $J(M)$; thus

$$P(M) = \underline{Z}J(M),\ \underline{Z}\underline{C}(M,-)\otimes P(M) = \underline{Z}(C(M,-)\times J(M)),$$

and finally

$$\tilde{J} = \tilde{P} = \coprod \underline{Z}(C(M,-)\times J(M)).$$

These $\underline{E}$-projective functors $\tilde{J}$ are exactly the "free" functors used by E.H. SPANIER in his presentation of the method of acyclic models ([13], p. 184), hence our comparison theorem includes his.

Finally, we show that there are enough $\underline{E}$-projectives in $\underline{A}^{\underline{C}}$. Let $F: C \longrightarrow A$ be any functor. For every model M let $p(M): P(M) \longrightarrow F(M)$ be an epimorphism with projective domain $P(M)$. These $p(M)$ exist since

$\underline{A}$ is assumed to have enough projectives. Let

$$\phi : \coprod Z\underline{C}(M,-) \otimes P(M) \longrightarrow F$$

be the morphism which corresponds to the family $(p(M))_{M \in \underline{M}}$ under the isomorphisms

$$\begin{aligned} &\underline{A}^{\underline{C}}(\coprod\{Z\underline{C}(M,-)\otimes P(M);\, M \in \underline{M}\},\, F) \\ &\qquad \cong \prod \{\underline{A}^{\underline{C}}(Z\underline{C}(M,-)\otimes P(M), F);\, M \in \underline{M}\} \\ &\qquad \cong \prod \{\underline{A}(P(M), F(M));\, M \in \underline{M}\}. \end{aligned}$$

The morphism ϕ is the unique one which makes the diagrams (i the injection, φ canonical)

$$\begin{array}{ccc} \coprod Z\underline{C}(M,C)\otimes P(M) & \xrightarrow{\quad \phi(C) \quad} & F(C) \\ \uparrow{\scriptstyle i \otimes 1} & & \\ Z\gamma \otimes P(N) & & \uparrow{\scriptstyle F(\gamma)} \\ \uparrow{\scriptstyle \varphi} & & \\ P(N) & \xrightarrow{\quad p(N) \quad} & F(N), \end{array}$$

for $N \in \underline{M}$, $C \in \underline{C}$, $\gamma \in \underline{C}(N,C)$, commutative. Also ϕ is in $\underline{E}$ since for $M \in \underline{M}$ we have

$$F(id_M)\, p(M) = \phi(M)\,(inj \otimes P(M))\varphi ,$$

and thus $\phi(M)$ is an epimorphism. Thus

$$\phi : \coprod Z\underline{C}(M,-)\otimes P(M) \longrightarrow F$$

is in $\underline{E}$ with $\underline{E}$-projective domain which means that there are enough $\underline{E}$-projectives in $\underline{A}^{\underline{C}}$. Since $\underline{E}$ satisfies the condition (PC) this implies that $\underline{E}$ is a projective class.

Moreover F is $\underline{E}$-projective if and only if the morphism

$$\phi: \coprod \underline{Z}\underline{C}(M,-)\otimes P(M) \longrightarrow F$$

has a right inverse, for any choice of $p(M)$ and $P(M)$. In particular, if $\underline{A} = \underline{Ab}$, one may choose $P(M)$ to be the free group $Z(|F(M)|)$ generated by the underlying set $|F(M)|$ of $F(M)$, with $p(M)$ the evident map. This is the choice made in [5], and it shows (in the case $\underline{A} = \underline{Ab}$) that F is $\underline{E}$-projective if and only if F is representable with respect to the model set $\underline{M}$ in the sense of Eilenberg-MacLane [4]. In particular, our new presentation of acyclic model theory gives at last a satisfactory explication of the notion of "representable" functor in the sense of EILENBERG-MACLANE, and also shows how this notion is connected with the more recent notion of representable functor $\underline{C}(M,-)$ in the sense of GROTHENDIECK (see e.g. [10]).

Remark 1. Theorem 2 can also be proved by using a theorem of EILENBERG-MOORE (Theorem 2.1 ff of [6]) which states that a pair of adjoint functors may be used to transfer a projective class from one category to another. Indeed, for each model $M \in \underline{M}$ the assignment $F \longrightarrow F(M)$ gives a functor

$$pr_M: \underline{A}^{\underline{C}} \longrightarrow \underline{A} \;;$$

by the isomorphism (2) it has a left adjoint $pr_M{}^*$ given by

$$pr_M{}^*(A) = \underline{Z}\underline{C}(M,-)\otimes A\,.$$

Now let $\underline{E}'$ be the projective class of all monomorphisms of A. By the cited transfer theorem, $pr_M^{-1}(\underline{E}')$ is a projective class in $\underline{A}^{\underline{C}}$, and hence so is the class

$$\underline{E} = \bigcap \{ pr_M^{-1}(\underline{E}') \; ; \; M \in \underline{M} \}$$

of the theorem. The same reference shows that the $\underline{E}$-projectives are exactly the retracts of direct sums of objects of the type $pr_M{}^*(P) = \underline{Z}\underline{C}(M, -) \otimes P$, with P projective in $\underline{A}$. This completes the proof.

Remark 2. Theorem 2 can be generalized as follows: For each model M, take an arbitrary projective class $\underline{E}_M$ in A, and define $\underline{E}$ to be the class of those natural transformations $e: F \longrightarrow G$ such that each $e(M)$ lies in the corresponding class $\underline{E}_M$. The theorem then asserts that $\underline{E}$ is a projective class, and that the $\underline{E}$-projectives are exactly the direct summands of direct sums of objects of the type $\underline{Z}\underline{C}(M, -) \otimes P(M)$, for any model M and any $\underline{E}'_M$-projective object $P(M)$. In particular, if for each M one chooses $\underline{E}_M$ to be the trivial projective class (all split epimorphisms), one finds that the $\underline{E}$-projectives are exactly the direct summands of direct sums of André's "elementary functors" (see [1]).

Theorem 2 can at once be dualized. Assume that $\underline{A}$ is an abelian category with infinite direct products, and let

$$\text{Hom}: \underline{Ab} \otimes \underline{A} \longrightarrow \underline{A}$$

be the formal Hom-functor (see [10]).

Theorem 2^*: Let $\underline{M}$ be a set of objects in a category $\underline{C}$, and let $\underline{A}$ be an abelian category with infinite direct products and enough injectives. In the category $\underline{A}^{\underline{C}}$ let $\underline{I}$ be the class of all those morphisms $i: F \longrightarrow G$ such that $i(M): F(M) \longrightarrow G(M)$ is a monomorphism for every model $M \in \underline{M}$. Then $\underline{I}$ is an injective class, and the $\underline{I}$-injectives are exactly the retracts of direct products of functors of the type $\mathrm{Hom}(\underline{Z}\underline{C}(-, M), I)$, $M \in \underline{M}$, I injective in $\underline{A}$.

A different form of the method of acyclic models has been used by M. BARR and J. BECK ([2], [3]) in their study of monads (this word is used to replace the inappropriate word "triple"). This is also included in the dual of our comparison theorem. We write I for the identity functor.

Theorem 3: Let $L: \underline{C} \longrightarrow \underline{C}$ be a functor and $\underline{A}$ an abelian category. In the abelian category $\underline{A}^{\underline{C}^o}$ of contravariant functors on $\underline{C}$ to $\underline{A}$, take $\underline{E}_L$ to be the class of those $e: F \longrightarrow G$ such that $eL: FL \longrightarrow GL$ is a split monomorphism; i.e., has a left inverse. Then the $\underline{E}_L$-injectives include the functors J for which there is a natural transformation $\varepsilon : L \longrightarrow I$ such that

$$J = JI \xrightarrow{J\varepsilon} JL$$

is a split monomorphism. Moreover, if either

i) $\underline{C}$ is small and $\underline{A}$ is complete, or

ii) L is the first component of a comonad (L, Δ, ε)

then $\underline{E}_L$ is an injective class, and in case (ii) the $\underline{E}_L$-injectives are exactly the retracts of functors HL, $H \in \underline{A}^{\underline{C}^o}$.

In the notation, eL is the natural transformation with $(eL)(C) = e(L(C))$ for every object $C \in \underline{C}$; similarly

$$(J\mathcal{E})(C) = J(\mathcal{E}(C)) \quad \text{for } J\mathcal{E} : J \longrightarrow JL .$$

Proof. Let $e: F \longrightarrow G$ be in $\underline{E}_L$ and $h: F \longrightarrow J$ be any morphisms. Let s and θ be left inverses of eL, resp. $J\mathcal{E}$, and define $h': G \longrightarrow J$ as the composite

$$G \xrightarrow{G\mathcal{E}} GL \xrightarrow{s} FL \xrightarrow{hL} JL \xrightarrow{\theta} J .$$

Then $h'e = h$, so J is $\underline{E}_L$-injective.

i) Under the assumptions of i) it is known that the functor

$$\underline{A}^{\underline{C}^o} \xrightarrow{\underline{A}^{L^o}} \underline{A}^{\underline{C}^o}$$

has a right adjoint L^o_* [8]. Let E' on $\underline{A}^{\underline{C}^o}$ be the trivial injective class consisting of all split monomorphisms. Every object of $\underline{A}^{\underline{C}^o}$ is $\underline{E}$-injective. Again by the Theorem 2.1 of [6] it follows that

$$(\underline{A}^{L^o})^{-1}(\underline{E}') = \{e ; (\underline{A}^{L^o})e = eL \text{ is a split monomorphism}\} = \underline{E}_L$$

is an injective class, and that the $\underline{E}_L$-injectives are exactly the retracts of objects

$$L^o_* H , \quad H \in \underline{A}^{\underline{C}^o} .$$

ii) Assume that $(L, \Delta, \mathcal{E})$ is a comonad. This consists of the functor L and natural transformations $\Delta: L \longrightarrow LL$ and $\mathcal{E} : L \longrightarrow I$ such that

Δ is associative and ε is a left and right counit of Δ. This latter means that both composites

$$L \xrightarrow{\Delta} L^2 \xrightarrow{L\varepsilon} L \quad \text{and} \quad L \xrightarrow{\Delta} L^2 \xrightarrow{\varepsilon L} L$$

are the identity. Hence for any contravariant functor $H \in \underline{A}^{\underline{C}^o}$ both composites

$$HL \xrightarrow{(HL)\varepsilon} (HL)L \xrightarrow{H\Delta} HL \quad \text{and} \quad HL \xrightarrow{(H\varepsilon)L} (HL)L \xrightarrow{H\Delta} HL$$

are the identity. The first implies that the morphism

$$HL \xrightarrow{(HL)\varepsilon} (HL)L$$

is a split monomorphism, and hence HL is $\underline{E}_L$-injective by the first part of the theorem. The second implies that the morphism

$$H \xrightarrow{H\varepsilon} HL$$

lies in $\underline{E}_L$ by definition. Hence there are enough $\underline{E}_L$-injectives in $\underline{A}$. Since $\underline{E}_L$ obviously satisfies the dual condition to (PC), $\underline{E}_L$ is closed and thus an injective class. Finally, H is $\underline{E}_L$-injective if and only if $H \xrightarrow{H\varepsilon} HL$ has a left inverse, or if and only if H is a retract of some FL, $F \in \underline{A}^{\underline{C}^o}$.

The arguments in this case are essentially a translation to the functor category of the considerations in EILENBERG-MOORE [6], p.391 .

BIBLIOGRAPHY

[1] ANDRÉ, M. Méthode simplicial en algèbre homologique et algèbre commutative, Springer-Verlag, Heidelberg (1967).

[2] BARR, M. Shukla Cohomology and Triples, Journal of Algebra 5 (1967).

[3] BARR, M. and BECK, J. Acyclic Models and Triples, Proceedings of the Conference on Categorical Algebra, La Jolla 1965 (Springer-Verlag, New York 1966), 336-343.

[4] DOLD, A. Ueber die Steenrodschen Kohomologieoperationen, Annals of Math. 73 (1961), 258-294.

[5] EILENBERG, S. and MACLANE, S. Acyclic Models, Amer. J. Math. 75 (1953), 189-199.

[6] EILENBERG, S. and MOORE, J. Foundations of Relative Homological Algebra, Memoirs Am. Math. Soc. Nr. 55, 1956.

[7] ____________, Adjoint Functors and Triples, Ill. J. Math. 9 (1965), 381-398.

[8] KAN, D. M. Adjoint Functors, Trans. Am. Math. Soc. 87 (1958), 294-329.

[9] MacLANE, S. Homology, Springer-Verlag, Heidelberg, 1963.

[10] MacLANE, S. Categorical Algebra, Bull. Am. Math. Soc. 71 (1965), 40-106.

[11] MARANDA, J. M. Injective Structures, Trans. A.M.S. 110 (1964), 98-135.

[12] MITCHELL, B. Theory of Categories, Academic Press, New York, 1965.

[13] SPANIER, E. H. Algebraic Topology, McGraw-Hill, New York, 1966.

EQUATIONAL SYSTEMS OF FUNCTORS*

Robert Davis, Southern Methodist University, Dallas, Texas

Introduction. Let $\underline{A}$ be any category. It is well known that there are two different elementary generalizations of classical universal algebra to a concept of algebraic structure on the objects of $\underline{A}$, namely the theories of monadic categories and of equational categories over $\underline{A}$. Specifically, by a monadic category we mean a triplable category in the sense of [1]; an equational category over $\underline{A}$, where $\underline{A}$ has enough products, is the category of product-preserving functors from $\underline{T}$ to $\underline{A}$, where $\underline{T}$ is an algebraic theory in the sense of Lawvere or of Linton (see [3]). For many categories $\underline{A}$, these two notions do not coincide. For example, if $\underline{A}$ has finite coproducts and a zero object, and A is a nonzero object of $\underline{A}$, then the category of objects below A is monadic but not equational since the forgetful functor does not send the initial object to zero. Again, if $\underline{A}$ is an ordered class, only $\underline{A}$ is equational over $\underline{A}$, whereas the categories monadic over $\underline{A}$ are the subcategories consisting of those elements which are closed under a fixed closure operation. For an example in the other direction, the category of finite groups is equational but not monadic over the category of finite sets.

---

*This paper represents part of the author's doctoral dissertation, Tulane University, New Orleans, 1967. This research was supported by a National Science Foundation Cooperative Graduate Fellowship at Tulane, and by ONR grant Nonr(G)-00040-66, NR 043-341 and NSF grant NSF GP-5609.

Nevertheless, monadic and equational categories have several important common structural properties, especially (1) the coequalizer condition in Beck's characterization of monadic categories (see [4]), and (2) the fact that if $\underline{A}$ is left complete so is each equational and each monadic category over $\underline{A}$, and the forgetful functor is limit-preserving. In this note we propose a common generalization of the two notions, and give a number of examples.

1. Definition. An equational system of functors $\underline{F} = (F_\alpha, \pi_{\alpha\gamma\delta}, \underline{C}, \underline{E})$ over $\underline{A}$ consists of a class of endofunctors F_α of $\underline{A}$, a class of natural transformations $\pi_{\alpha\gamma\delta}: F_\alpha \to F_\gamma$, a category $\underline{C}$ whose objects are the indices α, and a class $\underline{E}$ of equations whose permissible forms are described explicitly below.

The category $\underline{A}(\underline{F})$ of algebras in $\underline{A}$ over $\underline{F}$ is defined as follows. An algebra over $\underline{F}$ is a pair (A,X), where A is an object of $\underline{A}$ and X is a functor from $\underline{C}$ to $\underline{A}$ such that $X(\alpha) = F_\alpha(A)$. For each $\sigma: \alpha \to \gamma$ in $\underline{C}$, we shall write $X(\sigma) = \sigma_A: F_\alpha(A) \to F_\gamma(A)$, and by abuse of language we often say A is an algebra. The functor X is required to satisfy the equations in $\underline{E}$, each of which may be in either of two forms:

(1) If $\sigma: \alpha \to \gamma$, then $X(\sigma) = (\pi_{\alpha\gamma\delta})_A$ for some δ;

(2) $X(\sigma) = F_\lambda(X(\tau))$, where $\sigma: \alpha \to \gamma$, $\tau: \delta \to \zeta$, $F_\alpha = F_\lambda F_\delta$, and $F_\gamma = F_\lambda F_\zeta$.

A homomorphism $f: (A,X) \to (B,Y)$ is defined to be a map $f: A \to B$ in $\underline{A}$ such that for each $\sigma: \alpha \to \gamma$ in $\underline{C}$, the following diagram commutes:

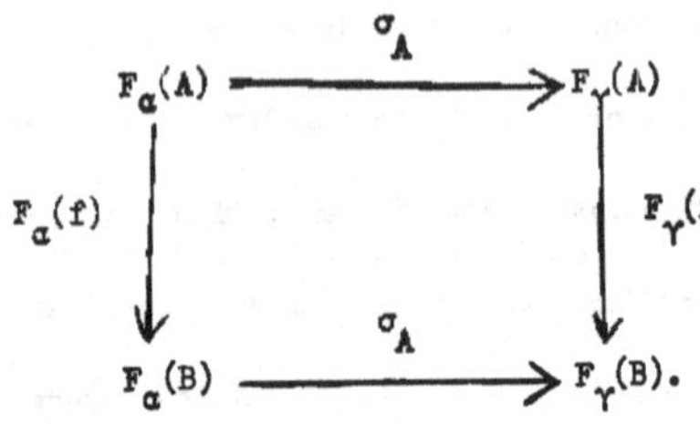

There is a forgetful functor $U: \underline{A}(\underline{F}) \to \underline{A}$ defined by $U(A,X) = A$ and $U(f) = f$. If $U': \underline{B} \to \underline{A}$, we shall say that $\underline{B}$ (or U') is of the form $\underline{A}(\underline{F})$ if there exists a system $\underline{F}$ and an equivalence $\underline{B} \to \underline{A}(\underline{F})$ which commutes with the functors to $\underline{A}$.

For example, let (T, η, μ) be a monad in $\underline{A}$. Then $\underline{A}^{(T)}$ is of the form $\underline{A}(\underline{F})$, where $\underline{F}$ is constructed as follows. Let $F_0 = I_{\underline{A}}$, $F_1 = T$, $F_2 = T^2$, let the π's be η, μ, and ηT, and let $\underline{C}$ be a category of the form

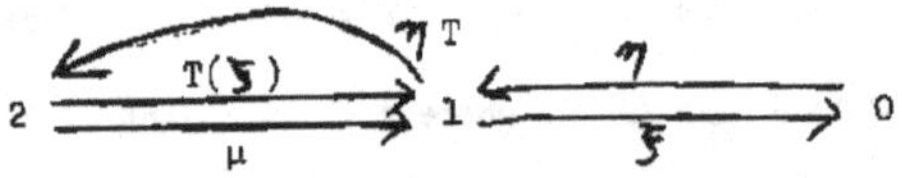

where $\xi\mu = \xi T(\xi)$, $\xi\eta = 0$, $\mu.\eta T = 1$, and $\eta\xi = T(\xi).\eta T$. Let the equations in $\underline{E}$, for an algebra (A,X), be $X(\eta) = \eta_A$, $X(\mu) = \mu_A$, $X(\eta T) = \eta_{T(A)}$, and $X(T(\xi)) = T(X(\xi))$. Then it is easy to see that $\underline{A}(\underline{F})$ is equivalent to $\underline{A}^{(T)}$.

One obtains equational categories in a similar manner, taking F_n to be the nth power functor, the π's to be the projections, $\underline{C}$ to be the equational theory under consideration, and $\underline{E}$ to consist of equations stating that X preserves projections. Other examples include comonadic and coequational categories, and the categories of machines considered in later sections.

For a further example, we can consider the generalized algebras defined in the thesis of Benabou. Let I be a set and M the free monoid generated by I. A *theory* of multiplicity I is a category $\underline{T}$ whose objects are the words $w = a_1 \dots a_n$ of M, where the a_i are elements of I, with the property that w is the product $a_1 x \dots x a_n$ in $\underline{T}$. If $\underline{A}$ has finite products, we define $\underline{A}^{(T)}$ to be the category of product-preserving functors from $\underline{T}$ to $\underline{A}$. There is a forgetful functor U: $\underline{A}^{(T)} \longrightarrow \underline{A}^I$ which takes X: $\underline{T} \longrightarrow \underline{A}$ to the I-tuple $(X(a))_{a \in I}$. We claim that $\underline{A}^{(T)}$ is of the form $\underline{A}^I(\underline{F})$. To see this, fix $b \in I$ and let t be the terminal object of $\underline{A}$. For each word $w = a_1 \dots a_n$, define F_w: $\underline{A}^I \longrightarrow \underline{A}^I$ to be the functor which takes each I-tuple $(A_a)_{a \in I}$ to that I-tuple which has t in all coordinates except the b'th, where the entry $A_{a_1} x \dots x A_{a_n}$ appears. The description of the desired $\underline{F}$ is completed by letting $\underline{C}$ be $\underline{T}$, the π's be the various projections, and $\underline{E}$ consist of the equations saying that the functor X preserves projections.

2. Generalized Coequalizer Conditions. It is an unsolved problem to give a structural characterization of the categories of the form $\underline{A}(\underline{F})$, similar to the characterization of monadic categories due to Beck. In this section we present the strongest known conditions on a functor U: $\underline{B} \longrightarrow \underline{A}$ which are necessary for it to be of the form $\underline{A}(\underline{F})$. Since the conditions consist of two sequences of generalizations of Beck's theorem, we state that theorem here for the sake of completeness. Specifically, U: $\underline{B} \longrightarrow \underline{A}$ is monadic if and only if it has a left

adjoint and the following condition holds. Suppose $d_0, d_1: B_1 \to B_0$ in $\underline{B}$, and suppose that in $\underline{A}$ there is a diagram

$$U(B_1) \underset{U(d_1)}{\overset{U(d_0)}{\rightrightarrows}} U(B_0) \overset{s}{\underset{z}{\leftrightarrows}} A, \qquad t: U(B_0) \to U(B_1)$$

such that $zU(d_0) = zU(d_1)$, $zs = A$, $sz = U(d_0)t$, and $U(d_1)t = U(B_0)$. It is easy to see that $z = \mathrm{coeq}(U(d_0), U(d_1))$ in $\underline{A}$. Then we require that d_0 and d_1 have a coequalizer in $\underline{B}$, say $q: B_0 \to B$, such that $U(q)$ is isomorphic to z; and also that if $r: B_0 \to C$ is such that $U(r)$ is isomorphic to z, then r is isomorphic to q.

The conditions considered in this section arise from generalizing the above diagram, first to n-skeletons of a simplicial object with a contracting homotopy, and second to n-skeletons of a simplicial object with a system of degeneracy maps.

Suppose first that $U: \underline{B} \to \underline{A}$ and $n > 0$. Let $d_i^{(n)}: B_n \to B_{n-1}$, $i = 0, \ldots, n$ be maps in $\underline{B}$, and suppose that in $\underline{A}$ there is a diagram

$$U(B_n) \underset{h_{n-1}}{\overset{\{f_i^{(n)}\} = \{U(d_i^{(n)})\}_{i=0}^{n}}{\rightleftarrows}} U(B_{n-1}) \underset{h_{n-2}}{\overset{\{f_i^{(n-1)}\}_{i=0}^{n-1}}{\rightleftarrows}} A_{n-2} \underset{h_{n-3}}{\overset{\{f_i^{(n-2)}\}_{i=0}^{n-2}}{\rightleftarrows}}$$

$$\cdots \underset{h_1}{\overset{\{f_i^{(2)}\}_{i=0}^{2}}{\rightleftarrows}} A_1 \underset{h_0}{\overset{\{f_i^{(1)}\}_{i=0}^{1}}{\rightleftarrows}} A_0 \underset{h_{-1}}{\overset{f^{(0)}}{\rightleftarrows}} A, \qquad \text{such that,}$$

writing A_{-1} occasionally for A, we have the equations

$$f_i^{(m)} f_j^{(m+1)} = f_j^{(m)} f_{i+1}^{(m+1)}, \quad j \leq i$$

and

$$f_i^{(m+1)} h_m = h_{m-1} f_i^{(m)}, \qquad i = 0,\ldots,m$$

$$f_{m+1}^{(m+1)} h_m = A_m, \qquad m = 0,\ldots,n-1,$$

and $f^{(0)} h_{-1} = A$. Note first that the following universal property holds. If $m < n$ and $g\colon A_m \longrightarrow C$ in $\underline{A}$ is such that $gf_0^{(m+1)} = \ldots = gf_{m+1}^{(m+1)}$, then there is a unique $k\colon A_{m-1} \longrightarrow C$ such that $kf_0^{(m)} = \ldots = kf_m^{(m)} = g$. (Specifically, $k = gh_{m-1}$.)

Then we say that U satisfies property B(n) if for every such diagram there exist in $\underline{B}$ objects B_i for $i = -1,\ldots,n-2$, and maps $d_i^{(m)}\colon B_m \longrightarrow B_{m-1}$ for $i = -0,\ldots,m$, and $m = 0,\ldots,n-1$, such that $U(B_m) = A_m$ and $U(d_i^{(m)}) = f_i^{(m)}$. These maps must satisfy the following conditions:

(1) $d_i^{(m)} d_j^{(m+1)} = d_j^{(m)} d_{i+1}^{(m+1)}$ for $j \leq i$;

(2) if $g\colon B_m \longrightarrow C$ in $\underline{B}$ is such that $gd_0^{(m+1)} = \ldots = gd_{m+1}^{(m+1)}$, then there is a unique $k\colon B_{m-1} \longrightarrow C$ such that $kd_i^{(m)} = g$;

(3) the preservation and reflection conditions on U analogous to those in Beck's characterization theorem.

Property $\overline{B}(n)$ is defined similarly, except that the single h_m is replaced by a collection $h_i^{(m)}\colon A_{m-1} \longrightarrow A_m$, $i = 0,\ldots,m$, satisfying the classical simplicial identities. The universal condition is modified to read: if $g\colon B_m \longrightarrow C$ is such that $gd_0^{(m+1)} = \dots = gd_{m+1}^{(m+1)}$, then for each $i = 0,\dots,m$, there is a unique $k_i\colon B_{m-1} \longrightarrow C$ such that $k_i d_i^{(m)} = g$.

Proposition 1. For each $n > 0$, $U\colon \underline{A}(\underline{F}) \longrightarrow \underline{A}$ satisfies properties $B(n)$ and $\bar{B}(n)$ and their duals.

The proof is a straightforward computation and is omitted. Note, however, that $B(1)$ is the condition of Beck's theorem; hence we have the

Corollary. $U\colon \underline{A}(\underline{F}) \longrightarrow \underline{A}$ is monadic (resp. comonadic) iff it has a left (resp. right) adjoint. Furthermore, $U\colon \underline{B} \longrightarrow \underline{A}$ with a left (resp. right) adjoint is of the form $\underline{A}(\underline{F})$ iff it is monadic (resp. comonadic).

We may conjecture that properties $B(n)$ and $\bar{B}(n)$ and their duals are sufficient for a category to be of the form $\underline{A}(\underline{F})$; however, it is not even known, for example, whether the category of torsion abelian groups is of the form $\underline{S}(\underline{F})$ over the category $\underline{S}$ of sets. The conditions seem to be satisfied by almost anything that could be called a category of algebras or coalgebras in some reasonable sense.

3. Congruences and Products. The question of the existence and preservation of (co)limits in $\underline{A}(\underline{F})$ seems to be rather difficult; it is sufficient but not necessary, for example, that they exist in $\underline{A}$ and be preserved by all the F_α. For further details see [2]. In this section we present a condition for the existence of products in categories $\underline{S}(\underline{F})$ which applies in particular to the categories $\underline{\text{Mac}}(M,N,\underline{S})$ of the later sections.

Suppose A is an algebra in $\underline{S}(\underline{F})$. Then by a "congruence relation" ρ on A ought to be meant an equivalence relation such that the natural map $f: A \longrightarrow B = A/\rho$ induces an algebra structure on B which makes f a homomorphism. This requirement means that for every $\sigma: \alpha \longrightarrow \gamma$ in $\underline{C}$, there exists a (unique) fill-in for the diagram

$$\begin{array}{ccc} F_\alpha(A) & \xrightarrow{\ \sigma_A\ } & F_\gamma(A) \\ {\scriptstyle F_\alpha(f)}\downarrow & & \downarrow{\scriptstyle F_\gamma(f)} \\ F_\alpha(B) & \xrightarrow{\ \sigma_B\ } & F_\gamma(B). \end{array}$$

(The uniqueness forces the passage from σ to σ_B to yield an algebra structure on B.) In turn this means that if a,b $F_\alpha(A)$, then $F_\alpha(f)(a) = F_\alpha(f)(b)$ implies $F_\gamma(f)\sigma_A(a) = F_\gamma(f)\sigma_A(b)$; for then $\sigma_B F_\alpha(f)(a)$ is well-defined to be $F_\gamma(f)\sigma_A(a)$. We can state this more concisely by defining the relation $F_\alpha(\rho)$ on $F_\alpha(A)$ by $(a,b) \in F_\alpha(\rho)$ iff $F_\alpha(f)(a) = F_\alpha(f)(b)$. Then we say ρ is a <u>congruence</u> <u>relation</u> on A iff for every $\sigma: \alpha \longrightarrow \gamma$ in $\underline{C}$, we have

$$F_\alpha(\rho) \subseteq (\sigma_A^{\,2})^{-1} F_\gamma(\rho).$$

<u>Definition</u>. We say a system $\underline{F}$ satisfies <u>condition</u> $\underline{P}$ (for product) if for every F_α in $\underline{F}$ and every indexed family $\{A_i\}$ of sets, the natural map $F_\alpha(\prod A_i) \longrightarrow \prod F_\alpha(A_i)$ is <u>monic</u>.

<u>Proposition</u> <u>2</u>. Suppose $\underline{F}$ satisfies condition P and that $\{\rho_i\}$ is a family of

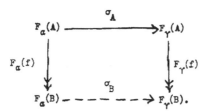

congruences on an algebra A in $\underline{S}(\underline{F})$. Then $\rho = \bigcap \rho_i$ is also a congruence on A.

Proof. The quotient map $f: A \to A/\rho$ is isomorphic to the coimage of the map $A \to \prod(A/\rho_i)$ taking $a \in A$ to the vector $(f_i(a))_i$, where $f_i : A \to A/\rho_i$ is the natural map. Let $B = A/\rho$, $B_i = A/\rho_i$, and let $k_\alpha : F_\alpha(\prod B_i) \rightarrowtail \prod F_\alpha(B_i)$ be the natural map. Then for each $\sigma: \alpha \to \gamma$, we have

$$\begin{aligned}
(x,y) \in F_\alpha(\rho) &\Rightarrow F_\alpha(f)(x) = F_\alpha(f)(y) \\
&\Rightarrow k_\alpha F_\alpha(f)(x) = k_\alpha F_\alpha(f)(y) \\
&\Rightarrow F_\alpha(f_i)(x) = F_\alpha(f_i)(y) \text{ for all } i \\
&\Rightarrow (x,y) \quad F_\alpha(\rho_i) \text{ for all } i \\
&\Rightarrow (x,y) \quad (\sigma_A{}^2)^{-1} F_\gamma(\rho_i) \text{ for all } i \\
&\Rightarrow F_\gamma(f_i)\sigma_A(x) = F_\gamma(f_i)\sigma_A(y) \text{ for all } i \\
&\Rightarrow k_\gamma F_\gamma(f)\sigma_A(x) = k_\gamma F_\gamma(f)\sigma_A(y) \\
&\Rightarrow F_\gamma(f)\sigma_A(x) = F_\gamma(f)\sigma_A(y) \text{ since } k_\gamma \text{ is monic} \\
&\Rightarrow (x,y) \in (\sigma_A{}^2)^{-1} F_\gamma(\rho). \quad \blacksquare
\end{aligned}$$

Proposition 3. Suppose $\underline{F}$ satisfies condition P and that either (1) $\underline{S}(\underline{F})$ has coproducts or (2) for every F_α and family $\{A_i\}$ of sets, the natural map $\bigcup F_\alpha(A_i) \to F_\alpha(\bigcup A_i)$ is onto. Then $\underline{S}(\underline{F})$ has products of nonempty families, and the product of a family of algebras A_i is an algebra structure on some subset of the set-product A_i.

Proof. Let $A = \prod A_i$ in $\underline{S}$, and suppose $f_i : D \longrightarrow A_i$ are homomorphisms. Let $k: D \twoheadrightarrow E$ be the coimage of the induced map $f: D \longrightarrow A$ in $\underline{S}$. Define the relation ρ on D by $(x,y) \in \rho$ iff $k(x) = k(y)$. Then ρ is an intersection of congruences and is therefore a congruence by proposition 2. Hence there is a unique algebra structure on E such that k is a homomorphism.

Next, suppose $f_i' : D' \longrightarrow A_i$, and define analogously the map $k' : D' \longrightarrow E' \subseteq A$. We claim that for each $\sigma: \alpha \longrightarrow \gamma$, σ_E and $\sigma_{E'}$ coincide on $F_\alpha(E) \cap F_\alpha(E') \supseteq F_\alpha(E \cap E')$. To see this, suppose $x \in F_\alpha(D)$, $y \in F_\alpha(D')$, and $F_\alpha(k)(x) = F_\alpha(k')(y)$. For each i, we have, where p_i is the projection to A_i,

$$\begin{aligned} F_\gamma(p_i)F_\gamma(k)\sigma_D(x) &= F_\gamma(f_i)\sigma_D(x) \\ &= \sigma_{A_i}F_\alpha(f_i)(x) \\ &= \sigma_{A_i}F_\alpha(p_i)F_\alpha(k)(x) \\ &= \sigma_{A_i}F_\alpha(p_i)F_\alpha(k')(y) \\ &= \sigma_{A_i}F_\alpha(f_i')(y) \\ &= F_\gamma(f_i')\sigma_{D'}(y) \\ &= F_\gamma(p_i)F_\gamma(k')\sigma_{D'}(y). \end{aligned}$$

Applying condition P, we have $\sigma_E F_\alpha(k)(x) = F_\gamma(k)\sigma_D(x) = F_\gamma(k')\sigma_{D'}(y) = \sigma_{E'}F_\alpha(k')(y)$, as desired.

The idea of the proof is that we can now glue together the algebra structures on all the E's thus obtained. Let $C \subseteq A$ be the union of all such subsets E. Each inclusion $E \longrightarrow C$ yields a monic $F_\alpha(E) \longrightarrow F_\alpha(C)$, so we can say that $\bigcup F_\alpha(E)$ is a subset of $F_\alpha(C)$. If hypothesis (1) holds, then each $x \in F_\alpha(C)$ is in $F_\alpha(E)$ for some E, and we put an algebra structure on C by defining $\sigma_C(x) = \sigma_E(x)$; the result just obtained shows that this value is independent of the choice of E.

Otherwise, for each E choose a D which yields E in the manner described above. Then $\coprod D$ admits a map to A which is $D \xrightarrow{k} E \longrightarrow A$ on each summand, and the image of this map is C. Thus, in the case of hypothesis (2), C is actually one of the E's.

We claim now that C is the product of the A_i in $\underline{S}(\underline{F})$. In fact, the only thing that is not immediately clear is that the projections $p_i | C$: $C \longrightarrow A_i$ are themselves homomorphisms. To see that such is the case, let $\sigma: \alpha \longrightarrow \gamma$ and $a \in F_\alpha(C)$. Then $a \in F_\alpha(E)$ for some E, and $a = F_\alpha(k)(x)$ for some x in the corresponding D. Suppose that $f_i: D \longrightarrow A_i$ are the homomorphisms by means of which E was obtained from D. Then

$$\begin{aligned}
\sigma_{A_i} F_\alpha(p_i)(a) &= \sigma_{A_i} F_\alpha(p_i) F_\alpha(k)(x) \\
&= \sigma_{A_i} F_\alpha(f_i)(x) \\
&= F_\gamma(f_i)\sigma_D(x) \\
&= F_\gamma(p_i) F_\gamma(k)\sigma_D(x)
\end{aligned}$$

$$= F_\gamma(p_i)\sigma_E F_\alpha(k)(x)$$

$$= F_\gamma(p_i)\sigma_C(a).$$

Hence $\sigma_{A_i} F_\alpha(p_i) = F_\gamma(p_i)\sigma_C$, and p_i is a homomorphism. ∎

4. Example: Categories of Machines. If M and N are monoids in $\underline{S}$, it is standard to define a sequential machine with input M and output N to be a set A of states together with functions $\delta: A \times M \longrightarrow A$, the transition function, and $\lambda: A \times M \longrightarrow N$, the output function, such that for all $a \in A$ and $m,n \in M$, we have $\delta(a,mn) = \delta(\delta(a,m),n)$, $\delta(a,0) = a$, and $\lambda(a,mn) = \lambda(a,m).\lambda(\delta(a,m),n)$. Then A is a right M-premodule, and we may write am for $\delta(a,m)$. We define the category $\underline{Mac}(M,N,\underline{S}) = \underline{Mac}(M,N) = \underline{Mac}$ of such machines by stipulating that a map from (A,δ_A,λ_A) to (B,δ_B,λ_B) should be a map f $\underline{S}(A,B)$ such that $f(am) = f(a)m$ and $\lambda_A(a,m) = \lambda_B(f(a),m)$. That is, the following diagram is to commute:

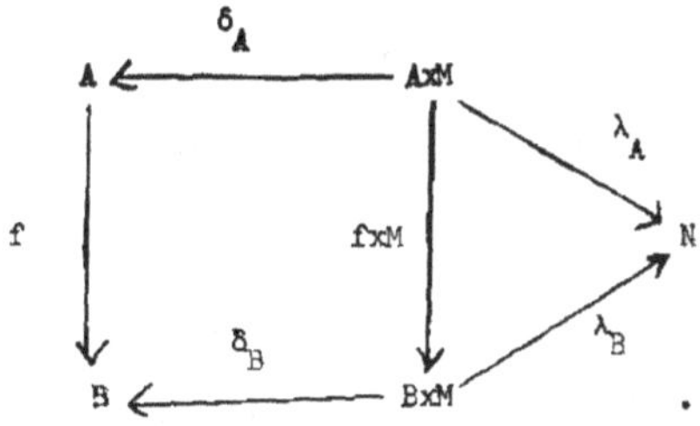

This definition can be immediately generalized to the concept of a category $\underline{Mac}(M,N,\underline{A})$, where $\underline{A}$ has finite products and a terminal object τ, as follows. M and N are to be monoids in $\underline{A}$, that is, objects of $\underline{A}$ with maps μ_M:

$M \times M \longrightarrow M$, $\mu_N : N \times N \longrightarrow N$, $t_M : \tau \longrightarrow M$, and $t_N : \tau \longrightarrow N$ satisfying the usual conditions. A machine in $\underline{A}$ with input M and output N is to be an object A of $\underline{A}$ with a map $\delta = \delta_A : A \times M \longrightarrow A$ making A a premodule over M, and another map $\lambda = \lambda_A : A \times M \longrightarrow N$ making the following diagram commute:

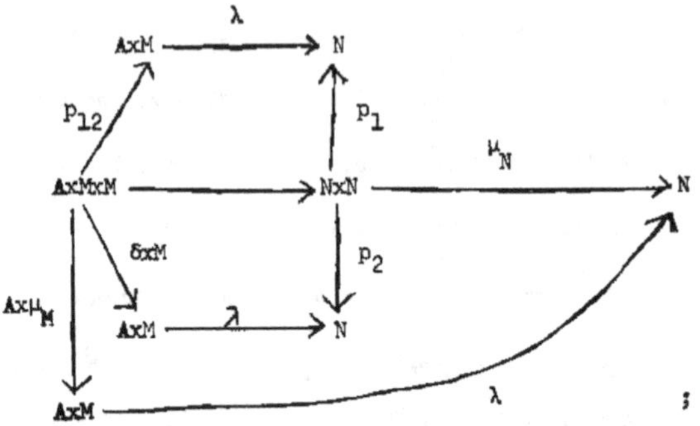

;

i.e., $\mu_N(\lambda p_{12}, \lambda(\delta \times M)) = \lambda(A \times \mu_M)$.

We retain the definition of morphism given above for the case $\underline{A} = \underline{S}$. There is an obvious forgetful functor $\underline{Mac}(M,N,\underline{A}) \longrightarrow \underline{A}$, and it is easy to see that $\underline{Mac}(M,N,\underline{A})$ is of the form $\underline{A}(\underline{F})$. For the remainder of this section we examine a few of these categories to indicate the variety of structure that can occur. For example, if $\underline{Ab}$ is the category of abelian groups, then it is easy to see that $\underline{Mac}(M,N,\underline{Ab})$ is equivalent to the disjoint union of copies of the category of groups below M, one copy for each element of $\underline{Ab}(M,N)$. Thus, in particular, if $\underline{Ab}(M,N) \neq 0$, then $\underline{Mac}(M,N,\underline{Ab})$ is not (co)monadic or (co)equational over $\underline{Ab}$.

The case $\underline{A} = \underline{S}$ is less trivial. Let us define $\underline{M}(M) = \underline{M}$ to be the category of right M-premodules, $U_1 : \underline{Mac}(M,N) \longrightarrow \underline{M}$ and $U_2 : \underline{M} \longrightarrow \underline{S}$ the forgetful functors, and $U = U_2 U_1$. The basic facts can be summed up in

Proposition 4. For any two monoids M and N, $\underline{Mac}(M,N,\underline{S})$ is left and right complete, well-powered and co-well-powered, and comonadic over $\underline{S}$.

Proof. It is straightforward to check that Mac is right complete and that U preserves colimits, and also that Mac has equalizers (which U also preserves!). Furthermore, by proposition 3 Mac has products of nonempty families.

Hence, to show Mac is left complete we need only show it has a terminal object; for this it now suffices to show that U satisfies the cosolution-set condition for 1. This condition means that there is a set $\{A_i\}$ of machines such that each machine A admits a map to one of the A_i. Let A be a machine and define the equivalence relation $\sim$ on A by $a \sim b$ iff $\lambda(a,-) = \lambda(b,-) : M \longrightarrow N$. Write $\bar{a}$ for the $\sim$-class of a. The number of $\sim$-classes in A is at most $\alpha = |N^M|$. For each $a \in A$ define $f_a : M \longrightarrow A/\sim$ by $f_a(m) = \overline{am}$, and define the equivalence relation $\approx$ on A by $a \approx b$ iff $f_a = f_b$. Then $\approx$ is a machine congruence on A, and the cardinal of $A/\approx$ is at most $\alpha^{|M|}$. Hence a maximal set of nonisomorphic machines of cardinal at most $\alpha^{|M|}$ provides a cosolution set for 1.

Let T be the terminal object of Mac. Using T, we can now show that the functor U_1 has a right adjoint, namely V_1: $\underline{M} \longrightarrow \underline{Mac}$ defined by $V_1(A) = A \times T$, with the product premodule structure $(a,x)m = (am,xm)$ and the output function $\lambda((a,x),m) = \lambda_T(x,m)$. Since it is well-known that U_2 has a right adjoint, we conclude that U also has a right adjoint.

Hence, in particular, U preserves epics; since it is also faithful, $\underline{Mac}$ is co-well-powered. To see that $\underline{Mac}$ is well-powered, suppose $f: A \longrightarrow B$ is in $\underline{Mac}$ and is not one-to-one. Suppose that $a,b \in A$ are such that $a \neq b$ but $f(a) = f(b)$, and define the functions $g,h: M \longrightarrow A$ by $g(1) = a$, $h(1) = b$. Then g and h are machine maps if we define an output on M by $\lambda_M(m,n) = \lambda_A(am,n) = \lambda_B(bm,n)$. But $g \neq h$, $fg = fh$ so f is not a monic in $\underline{Mac}$.

Finally, U is comonadic over $\underline{S}$ by the corollary to proposition 1. ∎

Remark 1. The collection of machine structures on M is a generating family for $\underline{Mac}$, and the right adjoint to U takes 2 to a cogenerator for $\underline{Mac}$. Hence any limit- (resp. colimit-) preserving functor with domain $\underline{Mac}$ must have a left (resp. right) adjoint. One such functor which might be worthy of investigation is the following. If A is in $\underline{Mac}(M,N)$ and B is in $\underline{Mac}(N,R)$, we can use the output of A as the input of B to obtain a machine $A \ominus B$ in $\underline{Mac}(M,R)$ called the series product of A and B. Precisely, $A \ominus B$ has underlying set $A \times B$, with $(a,b)m = (am, b.\lambda_A(a,m))$ and $\lambda((a,b),m) = \lambda_B(b,\lambda_A(a,m))$. The functors $- \ominus B$ and $A \ominus -$ are colimit-preserving; it is not clear what their right adjoints are.

Remark 2. The terminal object T of Mac is ordinarily rather large; for example, if M = N is a group, the underlying set of T can be shown to be the set of all maps in S(M,M) which take 1 to 1 (see [2]). The existence of a machine map f:A ⟶ B means intuitively that B can do anything that A can do, so T is a machine which has at least the same capability as any other machine in Mac(M,N,S).

Remark 3. One can also show that Mac is coequational over S (see [2]).

Remark 4. The above generalizations can be generalized slightly if we replace the monoids M and N in S by arbitrary small categories C and D. Then the category of M-premodules is replaced by the category of functors from C to S. If we fix a map $K : |C| \longrightarrow |D|$, where $|C|$ means the set of objects of C, then we can define an object of the category Mac(C,D,K) to be a functor $A: C \longrightarrow S$ equipped with an output function λ assigning to each $a \in A(C)$ and $f \in C(C,C')$ a map $\lambda(a,f): K(C) \longrightarrow K(C')$. Then each A(C) becomes a machine in the previous sense with input C(C,C) and output D(K(C),K(C)). Thus, we can think of A as a collection of sequential machines, connected in such a way that the input of a map $f: C \longrightarrow C'$ to machine A(C) at state a causes the collection to switch operation to the machine A(C') at state A(f)(a). The above results on Mac(M,N,S), except remark 3, extend to Mac (C,D,K) with only minor modifications.

It is of some interest to study categories Mac(M,N,A) where A is a category of topological spaces, partly because of the relationship with analog computers where output depends continuously on input. For example, suppose A = C is the category of compact spaces, and let M = N = R/Z be the circle group. We shall show that, in contrast to the case A = S, Mac(R/Z,R/Z,C) is not (co)monadic or (co)equational over C, nor does it have a cogenerator. It is fairly straightforward to show that the category is left complete and that the forgetful functor does not preserve products. Hence it suffices to prove the following result:

Proposition 5. If C is the category of compact T_2 spaces, the category Mac(R/Z,R/Z,C) is not right complete.

Proof. We exhibit a family of machines whose coproduct does not exist in Mac. Let A_i = R/Z for each $i > 0$, and define $\delta_i(a,m) = a+m$, and

$$\lambda_i(a,m) = f_i(a)-f_i(a+m),$$

where

$$f_i(x) = \min(1/2, ix(1-x)) \qquad \text{(in R)}.$$

Then A_i is a machine. Note that $f_i(0) = 0$ for all i, but if $x \neq 0$, then $\lim f_i(x) = 1/2$ as $i \to \infty$.

Suppose that the coproduct $A = \coprod A_i$ exists in Mac, with output function λ and injections $I_i : A_i \to A$; the injections must of course be machine maps. Let $a_i = I_i(0) \in A$. If the sequence $\{a_i\}$ assumes only a finite number of values, let a be a value assumed infinitely often. Otherwise let a be a cluster point of the sequence $\{a_i\}$ in the compact space A. Since λ is continuous, we have

$$\lambda(a,0) = \lim \lambda(a_i,0) = \lim \lambda_i(0,0) = 0,$$

but for $m \neq 0$,

$$\lambda(a,m) = \lim \lambda(a_i,m) = \lim \lambda_i(0,m) = 1/2.$$

Hence λ is not continuous at $(a,0)$, a contradiction. ▌

REFERENCES

1. Barr, M, and Beck, J., "Acyclic Models and Triples," Proceedings of the Conference on Categorical Algebra at La Jolla, Springer, 1967, 336-343.
2. Davis, R., "Abstract Universal Algebra," thesis, Tulane University, 1967.
3. Linton, F., "Some Aspects of Equational Categories," Proceedings of the Conference on Categorical Algebra at La Jolla, Springer, 1967, 84-94.
4. Manes, E., "A Triple Miscellany," thesis, Wesleyan University, 1967.

NORMAL COMPLETIONS OF CATEGORIES

John R. Isbell

Introduction

The normal completion theory is primarily for small categories. For them, a normal completion may be defined as a minimal complete extension, provided we are clear on the meanings of the terms. $\underline{E}$ is an extension of $\underline{C}$ when $\underline{C}$ is a full subcategory of $\underline{E}$; and in the same spirit, $\underline{E}$ is minimal among a class of extensions of $\underline{C}$ if $\underline{E}$ is a member of the class and not an extension of another member of the class. (Both of these terms and a number of others are often, and in this paper usually, applied more broadly "up to equivalence".) These usages are generally accepted, but my definition of complete is not. I call a category left small-complete if every small diagram has a limit; in most of the literature this is "left complete". For left complete I require, further, that every family of extremal subobjects of an object have an intersection.

In well-powered co-well-powered categories there is, of couse, no difference. Also, every normal completion of a small category is well- and co-well-powered. But it is unknown whether every minimal (left and right) small-complete extension is a normal completion. Here one must be careful with "(left and right)" (or it is not unknown). A normal completion of small $\underline{C}$ is both a minimal left small-complete and a minimal right small-complete extension of $\underline{C}$; the converse is true too; and the converse remains true if either occurrence of "minimal" is deleted. Thus non-normal minimal small-completions must be neither "left minimal" nor the dual; they must have other deficiencies, too.

The theory needs stronger completeness conditions than small-completeness even if it should turn out that every small-completion of a small category is normal. Aside from the theory (probably of limited application) of completing non-small categories, the Special Adjoint Functor Theorem is a principal tool, and it is now known [12] that small-completeness is insufficient for the form of the theorem that is needed.

The results on the normal completions $\underline{E}_\alpha$ of a small category $\underline{C}$ are these: First, $\underline{E}_\alpha$ being (as stated above) a minimal left small-complete extension, no proper subclass of the class of objects $|\underline{E}_\alpha|$ containing $|\underline{C}|$ is closed under formation of small limits; thus every object is a limit of limits transfinitely iterated from $|\underline{C}|$, and dually. There is at least one $\underline{E}_0$ all of whose objects are colimits of objects of $\underline{C}$ and limits of limits of objects of $\underline{C}$, and dually, but in general this is best possible (which answers negatively a question of Lambek [14]). Second, all $\underline{E}_\alpha$ at once can be embedded in a natural way in a category $\underline{Co}(\underline{C}, \underline{S})$ of "bimodules" over $\underline{C}$. Assuming appreciably less than the nonexistence of measurable cardinals, $\underline{Co}(\underline{C}, \underline{S})$ is fully embeddable in the category of algebras with one binary operation. Third, reflections and coreflections within $\underline{Co}(\underline{C}, \underline{S})$ establish a quasi-ordering of the completions $\underline{E}_\alpha$, whose equivalence classes form a complete lattice. The number of equivalence classes can be the power of the universe or perhaps more (not known), but the lattice is entirely complete. (Incidentally, the reflections and coreflections are faithful; every normal completion is embeddable, not fully, in every other.) Fourth, the left subregular (Yoneda) representation of any $\underline{\underline{E}}_\alpha$ over $\underline{\underline{C}}$ is faithful and reflects isomorphisms. (It is full for exactly one $\underline{\underline{E}}_0$.) Further properties, consequences of the first: the embedding $\underline{\underline{C}} \to \underline{\underline{E}}_\alpha$ preserves all limits that exist in $\underline{\underline{C}}$. $\underline{\underline{E}}_\alpha$ is cocompact, i.e., every functor on it that preserves limits (indeed, small limits) has an adjoint. (Dually, $\underline{\underline{E}}_\alpha$ is compact.) $\underline{\underline{E}}_\alpha$ is a retract of any category $\underline{\underline{F}}$ in which it is fully embedded; and $\underline{\underline{E}}_\alpha$ can be retracted upon any compact full subcategory by a reflector and a coreflector.

The "intersection" of the $\underline{\underline{E}}_\alpha$, the largest (up to equivalence) common subextension of $\underline{\underline{C}}$, can be rather simply described [11]. This does not seem to be true for the union. The intersection is, of course, incomplete except in the trivial case that all $\underline{\underline{E}}_\alpha$ are equivalent extensions. This seems likely to hold for the union too; at least, it need not be complete.

Practically everything carries over to normal completions of a category $\underline{\underline{C}}$ having small separating and coseparating subcategories, if one requires wide-completeness: every family of subobjects has an intersection, and dually. Of course "small" must be everywhere deleted or modified; and there is no result on representing $\underline{\underline{Co}}(\underline{\underline{C}}, \underline{\underline{S}})$ by algebras nor on a reflector-coreflector retraction. Some conditions are found under which these losses can be restored except for the representation by algebras; for instance, the category of CW-complexes and continuous functions satisfies such a condition.

Not much is proved in the most general normal completion theory, but it is formulated; one can hardly accept an arbitrary requirement of completeness in one case, wide-completeness in another. The general concept is normal extension $\underline{E} \supset \underline{C}$, meaning that no proper subclass of $|\underline{E}|$ containing $|\underline{C}|$ is closed under limit formation or colimit formation. Since this implies preservation of limits and colimits, there is no difficulty (in a strong set theory) in establishing maximal normal extensions. These are the completions already described, in the cases considered. On the other hand, the incomplete category of complete Boolean algebras is a maximal normal extension of itself. This generalizes: any equational category in Linton's sense [16] has the property that every small-valued limit functor is representable, which implies maximal normality.

The last few results in the paper (4.6 - 4.10) concern normal completions of the category of finite sets and other algebraic or varietal theories in the sense of [15] and [16].

The author is indebted to the National Science Foundation for support.

1. Normality (small case)

This section of the paper gives only known results, from [10]. Proofs are given, though sometimes informally and sometimes incompletely. We need not detail the set-theoretic foundation (modified Grothendieck universes in [10]), but the set theory should suffice for arbitrary well-orderings of the morphisms in any category considered. Ordinary Bernays-Gödel set theory is probably sufficient but certainly not convenient.

For this section, all categories $\underline{C}$ occurring have small hom sets $\underline{C}(X,Y)$.

Now a normal completion of $\underline{A}$ is not, except for introductory purposes, defined as a minimal complete extension of $\underline{A}$. It is defined as a complete extension $\underline{E}$ which is the full left closure of $\underline{A}$ and also the full right closure of $\underline{A}$. The full left closure is defined as the intersection of all containing, left closed, full subcategories, where left closed means closed under formation of small limits and intersections of arbitrarily many extremal subobjects.

The first basic results are the minimality of normal completions and the normality of minimal completions. The former depends mainly on the context-closure theorem of Lambek [14] and me [10], and the latter on Freyd's Special Adjoint Functor Theorem [1] as refined in [10]. The context-closure theorem says that "colimits respect limits", precisely as follows.

The full left context of a subcategory $\underline{A} \subset \underline{C}$ is the full subcategory of $\underline{C}$ on those objects B such that the principal ("hom") functor h^B: $\underline{A} \to \underline{S}$ preserves all limits. The theorem is that every colimit of such objects is such an object: the full left context is right universal. One can amplify; the full subcategory on the objects B such that h^B takes a given left conical extension (= left compatible family, of Freyd [1]) $\{f_\alpha: L \to D_\alpha\}$ of a diagram $\{D_\alpha; g_{\alpha\beta\nu}\}$ to a universal left conical extension is right universal. For if $\{j_\delta: B_\delta \to C\}$ is a universal right conical extension in $\underline{C}$, any left conical extension $\{r_\alpha: C \to D_\alpha\}$ induces left conical extensions $\{r_\alpha j_\delta$: all $\alpha\}$ for each δ. By hypothesis, for each δ there is exactly one $h_\delta: B_\delta \to L$ such that $f_\alpha h_\delta = r_\alpha j_\delta$. One checks easily that the h_δ form a right conical extension; so they are induced by a unique morphism $k: C \to L$, whence $f_\alpha kj_\delta = f_\alpha h_\delta = r_\alpha j_\delta$, and $f_\alpha k = r_\alpha$. Reversing the last calculation, k is unique.

Minimality is construed up to equivalence. A <u>total</u> subcategory is a full subcategory including an isomorph of every object. (In [10] I called this "dense". Ulmer has now introduced "dense" for a straightforward generalization of "left adequate", needlessly I think, but "dense" does seem too weak for the present "total".)

1.1 (from 3.7 of [10]). <u>In a normal completion $\underline{E}$ of $\underline{C}$, any left complete full subcategory containing $\underline{C}$ is total, and dually.</u>

<u>Proof</u>. A left complete full subcategory $\underline{D}$ containing $\underline{C}$ has full right closure $\underline{E}$, since $\underline{C}$ does. Hence the full left context of $\underline{D}$ is $\underline{E}$: every limit in $\underline{D}$ is a limit in $\underline{E}$ Hence every limit in $\underline{E}$ of a small diagram in $\underline{D}$ is isomorphic with an object of $\underline{D}$, and similarly for intersections of extremal subobjects. Thus the full subcategory $\underline{D}'$ on all isomorphs of objects of $\underline{D}$ is left closed. Since $\underline{E}$ is a normal completion of $\underline{C}$, $\underline{D}' = \underline{E}$, and $\underline{D}$ is total.

The rest of the cited result 3.7 [10] spells out the consequence of context-closure, that the embedding in a normal completion preserves all limits and colimits, and applies a constructive description of left and right closures. The fact is, a normal completion is precisely a generated and cogenerated completion in the sense of [10], which follows Grothendieck [4] and Semadeni [18]. But this terminology (like "complete") conflicts with the usual terminology. Since it matters for the Special Adjoint Functor Theorem, let us consider it now.

Grothendieck [4] calls an object G a generator in $\underline{C}$ if no object X has a proper subobject through which all morphisms from G to X factor, and defines <u>generating family</u> similarly. Of course this is a straightforward analogue of generating in algebra. In [4], in abelian categories, it is equivalent to the condition that any two distinct morphisms $X \to Y$ have distinct compositions with some morphism $G \to X$, which is weaker than algebraic generating even in commutative rings. Semadeni [18] repeats Grothendieck's definition, and adds the term <u>coseparate</u> for the morphism-distinguishing notion. Few authors follow. The usual practice is to attach the first term to the second notion. I am not acquainted with a reason for this. It is true that generating does not behave well in incomplete categories. In complete categories, one has 3.3.b of [10]:

<u>A right complete category $\underline{C}$ is generated by a subcategory $\underline{A}$ if and only if $\underline{C}$ is the full right closure of $\underline{A}$.</u>

In order to secure this equivalence for large $\underline{A}$, one must define carefully, but not unnaturally; $\underline{A}$ generates $\underline{C}$ if for each X in $|\underline{C}|$ there is a small set of objects of $\underline{A}$, not all of whose morphisms into X factor through any one proper subobject.

The proofs of this equivalence and of the next theorem rest on transfinite inductions. A reader who wants full details should go to

[10]; here they will be sketched. Concerning the relevance of these results to normal completions, the numbered one 1.2 is needed and the other is not.

1.2 (from 3.12 of [10]). _If_ $\underline{D}$ _is a left complete category cogenerated by a small subcategory, and_ $G: \underline{D} \to \underline{C}$ _preserves all limits, then_ G _has an adjoint._

The basic idea for these proofs is that of a _left multistrict analysis_: an inverse mapping system $\{X_\alpha; f_{\beta\alpha}\}$ indexed by an initial set of ordinals, in which each bonding morphism $f_{\alpha+1,\alpha}$ is a strict monomorphism (intersection of equalizers) and each X_ν with limit index is the inverse limit of its predecessors, with coordinate morphisms $f_{\nu\alpha}$. In the first place (and routinely; 2.3 of [10]), this implies that every $f_{\beta\alpha}$ is an extremal monomorphism. Second, in a left complete category, each morphism $g: W \to X_0$ factors through a finest left multistrict analysis; having $g = f_{\alpha 0}\, g_\alpha$, $f_{\alpha+1,\alpha}$ is a monomorphism representing the smallest strict subobject through which g_α factors. The analysis terminates, because of the strength of the set theory, and yields a _right image_ of g: a factorization $f_r e_r$, where e_r is epimorphic and represents the smallest quotient of W across which g factors (2.4 of [10]). Principal consequences: extremal monomorphisms are the same as multistrict monomorphisms (i.e. morphisms occurring in a left multistrict analysis), a left small-complete category is left complete if all left multistrict analyses have limits, and full left closures may be constructed by (1) adjoining small products, and (2) adjoining left multistrict analyses. (In (2), not _all_ analyses, only equalizers and intersections constructed from morphisms given or previously constructed;

see 3.3 of [10].) Secondary consequences: right images behave well, e.g. intersections of extremal monomorphisms are extremal.

Third, the full right closure of $\underline{A}$ is $\underline{C}$ if and only if $\underline{A}$ generates $\underline{C}$. Curiously, the proof (omitted in [10]) seems to require multistrict analyses both for "if" and for "only if". The class of objects generated by given objects A_λ includes the small coproducts C of the A_λ's, for a monomorphism $S \to C$ through which all morphisms $A_\lambda \to C$ factor would give (by coordinates) an inverse morphism $C \to S$. Similarly, one can go out a right multistrict analysis $C = Q_0 \to Q_1 \to \ldots$, for if all morphisms from A_λ's to Q_α factor through a subobject $S \to Q_\alpha$, so does $f_{0\alpha}: C \to Q_\alpha$, and so on. Conversely, if $\underline{A}$ generates $\underline{C}$, each X is the codomain of a morphism from a small coproduct C of A's which factors through no proper subobject; so $C \to X$ has left image $C \to X = X$, and is a multistrict epimorphism. In its (finest) multistrict analysis $\{X_\alpha; f_{\alpha\beta}\}$, each $f_{\alpha,\alpha+1}$ coequalizes a family of pairs of morphisms $p_{\alpha\lambda}, q_{\alpha\lambda}: W_{\alpha\lambda} \to X_\alpha$. Since each $W_{\alpha\lambda}$ is a multistrict quotient of a coproduct of A's, one can replace $(p_{\alpha\lambda}, q_{\alpha\lambda})$ with all pairs $(p_{\alpha\lambda}r, q_{\alpha\lambda}r)$, $r: A \to W_{\alpha\lambda}$, without changing the quotient.

Fourth, the Special Adjoint Functor Theorem 1.2. As in Freyd [1], one uses a small cogenerating subcategory $\underline{A}$ for any object X of $\underline{C}$ to construct a "universal" product P of A_λ's indexed by $\underline{C}(X, G(A_\lambda))$, and an evaluation $h: X \to G(P)$. There is an intersection $m: S \to P$ of all extremal monomorphisms n into P such that h factors through $G(n)$; since G preserves limits, $G(m)$ represents an intersection and h factors as $G(m)j$. Then $j: X \to G(S)$ represents $\underline{C}(X,G(\ ))$; for any y: X → G(Y) is G(v)j for a v: S → Y which is constructed (and unique, by induction) by backing out a left multistrict analysis Y → Q from some product Q of A's. Every $\underline{C}$(X,G( )) being representable, G has an adjoint.

The following result improves 3.19 of [10], where right small-completeness was assumed. That was an oversight; the assumption is removable by the result, two pages earlier in [10], that a left closure $\underline{B}$ of a small category is right complete. Slightly more is true. By 1.2, every limit-preserving functor on $\underline{B}$ to the category of small sets has an adjoint; so it is representable. By 1.1 and the Yoneda lemma, colimit functors are limit-preserving. Thus every small-valued colimit functor on $\underline{B}$ is representable, and this includes at least small coproducts and arbitrary cointersections. (The appeal to 1.1 is not necessary; it replaces an argument of Kan [13] used in [10].)

1.3. Every left complete extension of a small category $\underline{A}$ contains, fully, a normal completion of $\underline{A}$.

Proof. Let $\underline{B}$ be the full left closure of $\underline{A}$ in a left complete extension. As noted above, $\underline{B}$ is right complete; let $\underline{C}$ be the full right closure of $\underline{A}$ in $\underline{B}$. Again, $\underline{C}$ is left as well as right complete. The full right closure of $\underline{A}$ is $\underline{C}$, and since $\underline{A}$ cogenerates all of $\underline{B}$ it cogenerates $\underline{C}$: a normal completion.

(The last bit of the proof, as it is phrased here, requires the remark that the embedding $\underline{C} \subset \underline{B}$ preserves epimorphisms, by context-closure, and reflects isomorphisms, by fullness. It can be phrased otherwise so as to avoid the closure-generation equivalence, but I think not so as to avoid the bit of extra work.)

We have

1.4. A normal completion of a small category is the same as a minimal completion (up to equivalence) or a minimal left completion.

Moreover, each normal completion is a minimal left small-completion. This follows from the lemma (in 3.14 of [10]) that a category with finite limits and a small generating subcategory is well-powered. The lemma is easy and we omit it. It shows, as well, that a minimal right small-complete extension which is left small-complete is left complete; so it contains a normal completion; so it is a normal completion.

In [10], additional work is done on the lemma 3.14 to show (3.19.a) that if some small $\underline{A}$ has a minimal small-complete extension $\underline{E}$ that is not a normal completion, $\underline{A}$ is not separating or coseparating in $\underline{E}$. There is an example for the insufficiency of one-sided small-completeness, to appear in [12]; but no minimal one-sided small-complete extension is known (of small $\underline{A}$, of course) that is not complete.

2. Couples

The next results concern representations by sets and set-valued functors, so we had better specify the set theory. Let it be Zermelo-Fraenkel with the axiom of choice and with a distinguished inaccessible aleph, ∞. The point of ∞ is in the following definitions. A set of power $< \infty$, ∞, $> \infty$ respectively is called small, large, or extraordinary. A category is small if its set of morphisms is small, ordinary if it has at most ∞ objects and its hom sets are small, legitimate if its hom sets are small.

While the completion theory of categories with sets of objects can probably be done in Bernays-Gödel set theory without ∞, we shall soon be slightly further into the theory, where ∞ is a strict necessity. A simple example will illustrate. A category is called schlicht if each of its hom sets has at most one element. This term (of Hasse-Michler [5]) seems preferable to the common "preordered", but of course a schlicht category amounts to the same thing as a pre- or quasi-ordered class of objects with ordering " $\rightarrow$ ". Then consider the schlicht category of all subsets S of a set X of power ∞ such that S or X - S is small, ordered by inclusion. This is an ordinary category $\underline{C}$. Moreover, it has small limits and colimits, so that according to usual terminology it is complete. Still more, according to present terminology it is complete. Nevertheless it is obviously not "complete" in the sense of being all there. The normal completion theory for partially ordered sets is especially simple and conclusive; every partially ordered set has a unique normal completion, and the normal completion of $\underline{C}$ is the Boolean algebra $\underline{B}$ of all subsets of X. $\underline{B}$ has desirable properties that $\underline{C}$ lacks, such as compactness (defined in the Introduction: limit-preserving functors have adjoints). The only difficulty about $\underline{B}$ is its number of objects, which exceeds ∞. We shall admit it all the same.

(We do withhold the term "normal completion", calling $\underline{B}$ a maximal normal extension of $\underline{C}$; cf. Introduction.)

Now in order to state and prove three theorems on representations (one, 2.3, already stated in [10]) we must recall many definitions from [10] and earlier [7] and, for 2.7, from [11]. First, for a reader acquainted with [10].

"Sets" will be a systematically ambiguous symbol for any category whose objects are certain sets, whose morphisms are all the functions between those sets, in which there is an empty set and idempotents split. This is the "Zermelo-Fraenkel" substitute for the "category of all sets" of [10]. General hom functors can only be said to have values in Sets, which will not inconvenience us; conjugates of small-set-valued functors also are merely Sets-valued, and that will matter. $\underline{S}$ will mean some fixed category Sets of small sets which includes a set of each small power. We often imprecisely treat any small-set-valued functor as $\underline{S}$-valued.

A normal extension is defined only for a legitimate category $\underline{C}$, and is a legitimate extension which is the full left universal closure and the full right universal closure of $\underline{C}$. This and [10] suffice for 2.3 - 2.6.

A left universal (full) subcategory is one that is closed under taking limits, whatever limits exist; the full left universal closure is the smallest containing, full, left universal subcategory.

A conjugate of a functor $G: \underline{D} \to$ Sets is the following functor to another category Sets, or any naturally equivalent one. The functor is $G^*: \underline{D}^* \to \underline{Sets}$, each object $G^*(X)$ is the set of all natural transformations $\varphi: G \to h^X$, and each function $G^*(f)$ is the multiplication taking each $\varphi = \{\varphi_Z\}$ to $\{h_Z(f)\varphi_Z\}$. Note: the definition is applied both to covariant functors $G: \underline{C} \to \underline{Sets}$ and to contravariant functors $G: \underline{C}^* \to \underline{Sets}$ $(\underline{D} = C^*)$.

$G: \underline{C} \to \underline{Sets}$ (resp. $G: \underline{C} \to \underline{S}$) is a grounding (ordinary grounding) of $\underline{C}$. The dual is (ordinary) cogrounding. $\underline{Cat}(\underline{A},\underline{B})$ denotes the category of all functors from $\underline{A}$ to $\underline{B}$. The left regular (Yoneda) representation of a legitimate category $\underline{C}$, $\underline{C} \to \underline{Cat}(\underline{C}^*,\underline{S})$, is familiar; the left subregular representation over a subcategory $\underline{B}$ takes $\underline{C}$ to $\underline{Cat}(\underline{B}^*,\underline{S})$, each object C going to the restricted hom functor $h_C|\underline{B}^*$.

A coupling of two functors $G: \underline{C}^* \to \underline{Sets}$, $H: \underline{C} \to \underline{Sets}$ is a function m assigning to every ordered pair (p,q) with $p \in G(X)$, $q \in H(Y)$, a morphism $m(p,q): X \to Y$, satisfying $m(G(g)(p), H(h)(q)) = hm(p,q)g$ identically for $g: W \to X$, $h: Y \to Z$ in $\underline{C}$. An (ordinary) grounding couple F on $\underline{C}$ consists of a coupling m of an ordinary cogrounding $'F$ and an ordinary grounding F' of $\underline{C}$. This use of " ' " and the symbol " m " for any coupling will be usual. (We need no extraordinary couples.)

Morphisms $F \to J$ of grounding couples are conjoint transformations T, which are ordered pairs of natural transformation $'T: {}'F \to {}'J$, $T': J' \to F'$, satisfying $m('T_X(p),q) = m(p, T'_Y(q))$. With these morphisms we have a couple category $\underline{Co}(\underline{C},\underline{S})$. For legitimate $\underline{C}$, principal couples (h_X,h^X,m), coupled by $m(p,q) = qp$, yield the double regular representation $\underline{C} \to \underline{Co}(\underline{C},\underline{S})$. It is evident that the representation is full and faithful ([10]; 2.4 below). The left regular representation factors across it by Φ: Co(C,S) $\to$ Cat(C*,S), $\Phi(F) = {}'F$, $\Phi(T) = {}'T$. The double subregular representation over a full subcategory B takes X to $(h_X|B^*, h^X|B, m)$.

Couples Co(C,S) have two "forgetful functors", Φ taking each T to 'T and similarly $T \mapsto T'$ to Cat(C,S)*. It is Φ that acts like familiar forgetful functors; it preserves limits (which we shall not prove), and if C is small it has an adjoint. More generally, if the conjugate of every ordinary cogrounding of C is ordinary, then Φ has an adjoint Ψ, $\Psi(G) = (G, G^*, m)$, $m(p,q) = q_X(p)$ for $p \in G(X)$. The verification of adjointness is a straightforward application of the following correspondence, whose verification is also straightforward.

2.1 (1.1 of [10]). There is a one-to-one correspondence between couplings m of G with H and natural transformations $\mu: H \to G^*$, defined by

$$[\mu_Y(q)]_X(p) = m(p,q).$$

This formula is, naturally enough, the key to the algebra of couples. (Note that everything in 2.1 was defined at the ambiguous level of Sets.) It also mediates an adjunction not involving couples; for its (logically equivalent) dual identifies couplings m with natural transformations $'\mu: G \to H^*$. The adjoint functors, for small C, are the conjugations between Cat(C*,S) and Cat(C,S); they are contravariant and adjoint on the right. In connection with representations it is more natural to treat conjugations as between Cat(C*,S) and Cat(C,S)*. The functor K_2: Cat(C,S)* $\to$ Cat(C*,S) is defined by $K_2(H) = H^*$, $K_2(\Psi) = \Psi^*: H^* \to I^*$ where Ψ^*_X takes each $\alpha: H \to h^X$ to $\alpha\Psi$($\Psi: I \to H$ in Cat(C,S)). K_1^* is defined dually on Cat(C*,S) to Cat(C,S)*; K_1^* and K_2 are adjoint in that order, by routine check.

Further, the condition for two natural transformations $'T: 'F \to 'J$, $T': J' \to F'$ to make a conjoint transformation may be expressed in terms of the morphisms μ corresponding by 2.1 to the couplings m: $\mu T' = ('T)^*\mu$. (And dually.)

A couple F is separated if both $\mu: F' \to ('F)^*$ and the adjunct morphism $'\mu: 'F \to (F')^*$ are monomorphic. (A natural transformation π of groundings or any other functors is monomorphic if and only if each π_X is monomorphic; this is known and is quickly checked by using constant functors.)

We wish also to digress toward concreteness so far as to represent some couples by algebras. This requires (as far as is known) a limitation on the set theory, though not a bound. The assumption is

(A) For some small cardinal n, for every small cardinal m, every n-additive two-valued measure is m-additive.

(A) holds if (for instance) ∞ is isolated among the inaccessible cardinals.

2.2 Assuming (A), the following properties of a legitimate category $\underline{C}$ are equivalent.

(a) $\underline{C}$ is fully embeddable in a primitive category of (small) algebras.

(b) $\underline{C}$ is fully embeddable in the category of all (small) algebras with two unary operations.

(c) $\underline{C}$ is fully embeddable in a category having a small adequate subcategory.

The parentheses in "(small) algebras" don't mean that smallness is not necessary; rather, an algebra, topological space, or other ordinary structure is assumed to have a small ground set unless it is explicitly called illegitimate. 2.2 simply combines known results: (a) $\Rightarrow$ (b) from [6], (A) and (b) $\Rightarrow$ (c) from [11], (c) $\Rightarrow$ (a) from [9]. A category satisfying (a) is called boundable.

2.3 Theorem. The double subregular representation of a normal extension of a category $\underline{C}$ over $\underline{C}$ is full and faithful.

Proof. The proof will show that the part of the extension $\underline{E}$ on which the representation is full and faithful contains $\underline{C}$ and is "closed". Its containing $\underline{C}$ is a rather evident special case of the double Yoneda Lemma, which we may as well prove in general.

2.4 Yoneda Lemma. For a couple X and a principal couple $\bar{C} = (h_C, h^C, m)$ of $\underline{Co}(\underline{C},\underline{S})$, $\mathrm{Hom}(\bar{C},X)$ is $'X(C)$ and $\mathrm{Hom}(X,\bar{C})$ is $X'(C)$.

More generally:

2.5. Given $T,U\colon X \to Y$ in $\underline{Co}(\underline{C},\underline{S})$, if X is separated and $'T = {}'U$ then $T = U$.

Proof. X being separated, the coupling m of X corresponds by the formula of 2.1 to a monomorphism $\mu\colon X' \to ('X)^*$; and Y's coupling to some $\bar{\mu}\colon Y' \to ('Y)^*$. Conjointness of T means $\mu T' = ('T)^* \bar{\mu}$. Since μ is monomorphic, $'T$ determines T.

2.5 implies its own dual since $\underline{Co}(\underline{C}^*,\underline{S})$ is dual to $\underline{Co}(\underline{C},\underline{S})$. Then 2.4 follows from these and the ordinary Yoneda Lemma.

For 2.3, faithfulness follows from faithfulness of the left subregular representation $\rho\colon \underline{E} \to \underline{Co}(\underline{C},\underline{S}) \to \underline{Cat}(\underline{C}^*,\underline{S})$, which holds because the set of all ordered pairs of objects (X,Y) of $\underline{E}$ such that ρ is one-to-one on $\underline{E}(X,Y)$ contains $|\underline{C}| \times |\underline{C}|$ and is closed under taking limits of Y's and colimits of X's.

From faithfulness of ρ and the dual, the double subregular representation σ takes $\underline{E}$ to separated couples.

Now the set of all (X,Y) such that $(*)$ σ takes $\underline{E}(X,Y)$ onto $\mathrm{Hom}(\sigma(X), \sigma(Y))$ contains $|\underline{C}| \times |\underline{C}|$, by 2.4. Passing to a colimit X of a diagram D of X_α's such that $(*)$ holds for (X_α, Y), consider any conjoint $T: \sigma(X) \to \sigma(Y)$. The coordinate morphisms $i_\alpha: X_\alpha \to X$ induce $\sigma(i_\sigma): \sigma(X_\alpha) \to \sigma(X)$ and thus $T_\alpha = T\sigma(i_\alpha): \sigma(X_\alpha) \to \sigma(Y)$. By $(*)$ and faithfulness, each T_α is $\sigma(t_\alpha)$ for a unique $t_\alpha: X_\alpha \to Y$. For the morphisms $s: X_\alpha \to X_\beta$ in D, $i\beta s = i_\alpha$, $T_\alpha = T_\beta \sigma(s) = \sigma(t_\beta s)$; hence $t_\beta s = t_\alpha$. So $\{t_\alpha\}$ determines $t: X \to Y$, $ti_\alpha = t_\alpha$. Consequently $\sigma(t)': h^Y|\underline{C} \to h^X|\underline{C}$ has the same coordinates in the $h^{X_\alpha}|\underline{C}$ as T'; $\sigma(t)' = T'$; by the dual of 2.5, $\sigma(t) = T$. Closure under limits of Y's follows by a similar argument.

2.6 Theorem. <u>The left subregular representation of a normal extension of a category $\underline{C}$ over $\underline{C}$ is faithful and reflects isomorphisms.</u>

<u>Proof</u>. We proved in 2.3 that this representation is faithful. Also, by 2.3, we may assume that the extension is already fully embedded in $\underline{Co}(\underline{C},\underline{S})$. Then consider any morphism $T: X \to Y$ left represented by an isomorphism $'T$. By double Yoneda, every morphism from a principal couple W to Y factors uniquely through T. The set of objects W for which that is true is right universal, for a morphism from a colimit of W_α's to Y induces morphisms from the W_α's, which factor through T, and uniqueness assures that this makes a right conical extension of the diagram which gives the required morphism from the colimit to X (unique since the $W_\alpha \to X$ determine it). We conclude that Y belongs to that set; T has a unique right inverse S. As $'S$ must be the inverse of $'T$, $'(ST) = '1$, and by 2.5 $ST = 1$.

Next we wish to show that $\underline{Co}(\underline{C},\underline{S})$ is boundable for small $\underline{C}$ if the hypothesis (A) holds. In a sense, this result is incapable of generalization, since a boundable category is by definition fully embeddable in something obviously fully embeddable in a $\underline{Co}(\underline{C},\underline{S})$. However, a more general formulation will be more natural and will help to isolate the role of hypothesis (A).

For any adjoint pair of functors $F: \underline{C} \to \underline{D}$, $G: \underline{D} \to \underline{C}$, let an (F,G) couple be a pair (X,Y) of objects, X in $\underline{C}$, Y in $\underline{D}$, with an adjunct pair of morphisms $'m: X \to G(Y)$, $m': F(X) \to Y$. A conjoint transformation $\alpha: (X,Y,m) \to (X_1,Y_1,m_1)$ is a pair $'\alpha: X \to X_1$, $\alpha': Y \to Y_1$, satisfying $'m_1 '\alpha = G(\alpha')'m$, or equivalently (adjointly) $m_1' F('\alpha) = \alpha'm'$. The (F,G) couples and conjoint transformations form a category $\underline{Cyl}(F,G)$. To see $\underline{Co}(\underline{A},\underline{S})$ as an instance (small $\underline{A}$), take $F = K_1^*: \underline{Cat}(\underline{A}^*,\underline{S}) \to \underline{Cat}(\underline{A},\underline{S})^*$, $G = K_2$; $'m$ and m' are just the $'\mu$ and μ associated (reversibly) with a grounding couple by 2.1.

2.7 Theorem. If $F: \underline{C} \to \underline{D}$, $G: \underline{D} \to \underline{C}$ are adjoint functors between boundable categories and $\underline{C}$ has a small coseparating subcategory, $\underline{Cyl}(F,G)$ is boundable.

Proof. The small coseparating subcategory $\underline{R}$ in $\underline{C}$ will fit into a construction from [9]. Given a left adequate full subcategory $\underline{Q}$ of a legitimate category $\underline{E}$, first replicate objects if necessary so that every object of $\underline{Q}$ is the codomain of a morphism whose domain is a different object of $\underline{Q}$, and then represent each object X of $\underline{E}$ by the algebra whose ground set is $U[\underline{E}(Q,X): Q \in |\underline{Q}|]$ and whose operations are unary operations $\bar{q}(f) = fq$ when this is meaningful, f otherwise. With the obvious representation of morphisms, this is full and faithful (4.2 of [9]).

Since $\underline{C}$ is boundable, it is fully embeddable in a category of unary algebras $\underline{E}$ whose free algebra on one generator, E, is left adequate. Apply the construction just described, using the full subcategory on $|\underline{R}|$ and E. Without further notation, we shall identify objects X of $\underline{C}$ with the corresponding algebras; thus $\cup\{Hom(R,X): R \in |\underline{R}|\}$ is a subset rX of X. Do the same with $\underline{D}$, using $F(\underline{R})$.

We define an algebra $A(X,Y,m)$ for each (F,G) couple (X,Y,m) as follows. The ground set is a disjoint union of the three sets $X,Y,F(X)$ and four additional points $0,1,2,3$. The latter points are the values of four 0-ary operations. One unary operation π takes X to 0, Y to 1, F(X) to 2, and $\{0,1,2,3\}$ to 3. Each operation of $\underline{C}$ is defined on X and extended over the rest of $A(X,Y,m)$ to coincide there with the identity. Similarly each operation of $\underline{D}$ is defined on $Y \cup F(X)$ and extended identically. There are two more unary operations: $\bar{m}$ takes F(X) to Y by m' and is extended identically; $\bar{F}$ takes rX into F(X) by $\bar{F}(f) = F(f)$ and is extended identically.

For any conjoint transformation $\alpha: (X,Y,m) \to (X_1,Y_1,m_1)$, there is a homomorphism $A(\alpha)$ defined on $A(X,Y,m)$ as $'\alpha$ on X, α' on Y, $F('\alpha)$ on F(X), $i \longmapsto i$ on $\{0,1,2,3\}$. ($A(\alpha)$ preserves $\bar{m}$ by conjointness, preserves $\bar{F}$ since F is a functor, and obviously preserves the other operations too.) We have a functor, and it is faithful since $A(\alpha)$ determines $'\alpha$ and α'. For fullness, any homomorphism $A(X,Y,m) \to A(X_1,Y_1,m_1)$ breaks up into three homomorphisms $'\alpha: X \to X_1$, $\alpha': Y \to Y_1$, $\beta: F(X) \to F(X_1)$, because of the operations $0,1,2,3,\pi$ and the operations of $\underline{C}$ and $\underline{D}$. Preservation of $\bar{m}$ means $m_1{}'\beta = \alpha' m'$. Finally, if β and $F('\alpha): F(X) \to F(X_1)$ were different, so would be their adjuncts $X \to GF(X_1)$ in $\underline{C}$. Since $\underline{R}$ is coseparating, some $f\colon R \to X$ must distinguish them, which is absurd since a homomorphism preserves $\tilde{F}$.

2.8 Corollary. Assuming (A), if C is a small category then $\underline{Co}(\underline{C},\underline{S})$ and every normal completion of C are boundable.

Proof. We are coupling $\underline{Cat}(\underline{C}^*,\underline{S})$, in which $\underline{C}$ is left adequate and hence coseparating, with $\underline{Cat}(\underline{C},\underline{S})^*$, in which $\underline{C}$ is right adequate. On the assumption (A), boundability is self-dual (2.2), so the theorem applies.

As an indication of the power of 2.8, topological spaces are couples over a two-point set (T_0 space) with its identity and constant functions. To continue not irrelevantly, $\underline{Cyl}(F,G)$ would seem to be of interest for general adjoint functors; the trouble is that it is too big. "$\underline{Cyl}$" for "cylinder" is not unreasonable. $\underline{C}$ is fully embedded by $C \mapsto (C,F(C),m)$, where $m' = 1$, and it is at the end of the cylinder in the sense that it is a coreflective subcategory. This generalizes half of 2.4. (That seems to be near the end of the results in this paper generalizing to $\underline{Cyl}(F,G)$. It is fundamental that we have an incomplete $\underline{A}$, not "closed" in $\underline{Co}(\underline{A},\underline{S})$.)

3. Maximal normality

This section contains nothing new on completions of small categories (except the "absolute retract" property, which is in 4.8 of [10] already); it treats generalizations of the results of Section 1. Some things implicit in Section 1 will receive explicit attention. For instance, "all limits" in 1.2 becomes "small limits" for a completion $\underline{E}$. This is because the limits involved in the proof are no worse than left multi-strict analyses, and those must be small since $\underline{E}$ is a generated completion and therefore well-powered.

3.1. A normal extension of a normal extension is normal.

Proof. If $\underline{C} \subset \underline{D} \subset \underline{E}$ normally, both embeddings are conservative (preserve all limits and colimits) by context-closure. So the full left universal closure of $\underline{C}$ contains $\underline{D}$ and therefore $\underline{E}$, and dually.

A legitimate category is maximal normal if it is total in all its normal extensions. A maximal normal normal extension will be called a maximal normal extension.

3.2 Every legitimate category has a maximal normal extension.

Proof. By 3.1, one can extend stepwise; because of context-closure, the union of a chain of normal extensions is normal; by 2.3, there is a bound $\underline{Co}(\underline{C},\underline{S})$ to secure termination.

It seems worth remarking that if $\underline{C} \subset \underline{D} \subset \underline{E}$ and $\underline{E}$ is normal over $\underline{C}$, then $\underline{E}$ is normal over $\underline{D}$, but $\underline{D}$ need not be normal over $\underline{C}$. Another side remark: Lambek has suggested [14] that in the best extensions $\underline{E}$ of $\underline{C}$, every object should be a limit of objects of $\underline{C}$ and a colimit of objects of $\underline{C}$. It is shown in [12] that the one-object category Z_4 has no such extension having finite limits. One can come close to completing a general small category $\underline{C}$ in the way Lambek suggested; it was shown in [10] that the full left closure of $\underline{C} \subset \underline{Cat}(\underline{C}^*,\underline{S})$ is a completion in which every object (stable functor) is a colimit, and a limit of limits, of objects of $\underline{C}$. The completion might be called a Lambek extension of a Lambek extension.

As stated in the Introduction, we shall call a legitimate category $\underline{C}$ <u>compact</u> if every functor from $\underline{C}$ to a legitimate category preserving all colimits has a coadjoint. Equivalently, every limit-preserving functor $\underline{C}^* \to \underline{S}$ is representable. We attach the term "compact" to this property, <u>cocompact</u> to the dual, because this is the property that the category of groups has. Indeed, every primitive category of algebras is compact, by the dual of 1.2. On the category of groups, the coproduct functor of a maximal set of pairwise non-isomorphic simple groups is small-valued (since for any group G, almost all simple groups S are not embeddable in G and thus Hom(S,G) = 0), and, like all colimit functors, it preserves limits; but it is not representable.

We avoid "left" and "right" in naming compactness because it tends to be deduced from colimits (as in the dual of 1.2) but what it implies, is limits. <u>Every small-valued limit functor on a compact category is representable</u>.

Undoubtedly some readers would prefer to avoid large diagrams, and define "compactness" by representability of contravariant functors (contra-) preserving small colimits. That is a pleasant property for a category to have; it is closely related to compactness plus chain conditions, cf. 3.10.

A category is called <u>left wide-complete</u> if it is left small-complete and every family of subobjects of an object has a (representable) intersection.

3.3 (from 3.12 of [10]). A left wide-complete legitimate category separated by a small subcategory is cocompact.

For the proof, similar to the proof of 1.2, we refer to [10]. Some details of 3.12 of [10] not stated in 1.2 and 3.3 are used in 3.10 below; the point is that functors preserving certain limits (small limits and appropriate intersections) are representable.

We could now go directly (through some lemmas) to the proof of most of Theorem 3.5, recalling one more definition from [10]: a grounding F is separated if the natural couple (F,F^*,m) is separated.

Here, at the place in this paper where serious use of separation begins, let us correct a non-optimal definition in [10]. Call a subcategory $\underline{A}$ of $\underline{B}$ separating if (only) for every two distinct morphisms $f, g: X \to Y$ there exist Z in $|\underline{A}|$ and $h: Y \to Z$ such that $hf \neq hg$. This removes a smallness condition (like the condition we retain for generating) and invalidates the lemma 3.3c of [10] except for small subcategories. Every use of separation or coseparation in [10] (not counting the lemma) is unaffected; the subcategories concerned are small except in 3.17, and both versions of 3.17 are true.

Recall further that a category is called extraordinary-injective if it is legitimate and is a retract of every legitimate extension. It is injective if it is legitimate and a retract of every legitimate extension having just one more object. That implies retractability of extensions having ∞ more objects (4.5 of [10]). It is not known whether injective $\Rightarrow$ extraordinary-injective.

Among so many sides and aspects of completeness, we should reduce the confusion by noting:

3.4 Every injective, or compact, or cocompact category is maximal normal.

The proofs are very brief applications of a theorem misstated in [10]. The statement is that given a retraction E → C upon a subcategory (not necessarily full), every diagram in C having a limit in E has a limit in C. I do not know whether this is true. The proof in [10] is sound if we add the hypothesis that every idempotent morphism in C splits in E. In the applications in [10] (4.3 and 4.9), that hypothesis is satisfied.

Proof of 3.4. If C is injective or compact or cocompact, idempotents split in C (routine; but we only need the first). Suppose C is injective and E a normal extension. For any limit X of a diagram in C, the full subcategory on C and X retracts upon C; so the diagram had a limit Y relatively in C. By context-closure, X and Y are isomorphic. Thus C is total in its left universal closure, and C is total in E. Second, suppose C is compact and E a normal extension. By context-closure, C → E preserves colimits; by compactness, there is an adjoint. Since C is full in E, there is an adjoint retraction; this is rather well known and is in 3.16 of [10], which adds the consequence that C is total in its right universal closure. Hence C is total in E. Duality concludes it.

3.5 Theorem. A maximal normal category having small separating and coseparating subcategories is compact, cocompact, and extraordinary-injective.

The proof of compactness can be sketched quickly. It will suffice to establish left wide-completeness; for then 3.3 yields cocompactness, whence every small-valued colimit functor is representable, and compactness follows from the dual of 3.3. Now if we adjoin to the given category $\underline{C}$, in $\underline{\text{Cat}}(\underline{C}^*,\underline{S})$, any of the limit functors F whose representability is required for left wide-completeness, we have a Lambek extension $\underline{D}$ of $\underline{C}$. $\underline{D}$ is normal if it is legitimate. It remains to note that $\underline{C}$'s small coseparating subcategory makes $\underline{D}$ legitimate. This principle is not new, having been used at least by D. B. Fuks in [3]. We formulate it in three lemmas.

3.6. A category $\underline{D}$ is legitimate if it has a small separating subcategory $\underline{A}$ and $\underline{D}(X,A)$ is small whenever A is in $\underline{A}$.

Proof. Morphisms in $\underline{D}(X,Y)$ are determined by the multiplications they induce on all $\underline{D}(Y,A)$ to $\underline{D}(X,A)$, small families of small sets.

There is not any application to general extensions of $\underline{C}$, because separating is not transitive. Moreover, although $\underline{S}$ has a separator and separates $\underline{\text{Cat}}(\underline{S},\underline{S})^*$, $\underline{\text{Cat}}(\underline{S},\underline{S})$ is not legitimate; almost all of the extraordinary set of different functors constructed in [8] (in ordinary Bernays-Gödel set theory) have extraordinary sets of natural transformations to themselves. We need:

3.7. A separating subcategory of a separating and coseparating full subcategory of D is separating in D.

Proof. If the categories are $\underline{A} \subset \underline{B} \subset \underline{D}$, consider any distinct coterminal morphisms $x, y: X \to Y$ in $\underline{D}$. There is $w: W \to X$, $W \in |\underline{B}|$, such that $xw \neq yw$. Then there is $z: Y \to Z$, $Z \in |\underline{B}|$, such that $zxw \neq zyw$. Since $\underline{B}$ is full, zxw and zyw are in it, so that $azxw \neq azyw$ for some $a: Z \to A$, and $(az)x \neq (az)y$.

3.8 The conjugate of an ordinary grounding of a legitimate category B with a small separating subcategory is small-valued, and the full

<u>subcategory of Cat($\underline{B}$,$\underline{S}$) composed of the separated functors is legitimate</u>.

This follows at once from 3.7 and concludes the proof of compactness and cocompactness in 3.5. Then any legitimate extension of $\underline{C}$ may be retracted, by left subregular representation into Cat($\underline{C}$*,$\underline{S}$) followed by reflection into stable (equivalently, principal) cogroundings. Every small-valued stable functor is representable by compactness, and the stable reflection, explicitly described on pages 646-7 of [10], is small-valued by 3.8 and its dual. 3.5 is proved.

Three by-products of the proof follow.

3.5.a. In 3.5, "maximal normal" may be replaced by "left wide-complete".

3.5.b. A maximal normal category with a small separating subcategory is right wide-complete; indeed, every small-valued colimit functor is representable.

3.5.c. A normal extension of a category with a (small) separating subcategory has the same separating subcategory.

Conversely:

3.5.d. <u>If a category $\underline{C}$ has no small full separating subcategory, neither does any extension in which $\underline{C}$ is separating</u>.

<u>Proof</u>. Given an extension $\underline{E}$ in which $\underline{C}$ is separating and $\underline{A}$ is small, full, and separating, consider the set of all morphisms $A \to A'$ factoring across objects C. It is small, so it factors across a small number of C_i's. Any two distinct coterminal morphisms $X \to Y$ in $\underline{E}$ are distinguished by some morphism $Y \to A \to C \to A'$, and thus by some morphism $Y \to C_i$.

I do not know if 1.3 generalizes to this setting, i.e. if every left wide-complete extension contains a maximal normal extension. There is a precise analogue of the result on small-complete extensions stated after 1.4, as follows.

3.9. If $\underline{C}$ is legitimate and has a small separating subcategory, then any minimal right wide-complete extension of $\underline{C}$ that is left wide-complete and legitimate is a maximal normal extension.

Proof. In such an extension $\underline{E}$, for every non-invertible monomorphism $X \to Y$, there is a morphism $C \to Y$ not factoring through X. If this were false for some $m: X \to Y$, the set of objects W such that every morphism $W \to Y$ factors through m would contain $\underline{C}$ and be closed under the formation of colimits. (For, since m is monomorphic, factoring through it is a single-valued partial operation and takes right conical extensions to right conical extensions.

Let $\underline{A}$ be a small separating subcategory of $\underline{C}$ and $\underline{D}$ the full subcategory of $\underline{E}$ on all subobjects of small products of objects of $\underline{A}$. $\underline{D}$ contains $\underline{C}$. For, each object C of $\underline{C}$ has a universal morphism h to a small product of A's, whose coordinates are all $C \to A$. This h is monomorphic because, given $f, f: Y \to C$, $hf = hg$ implies that every morphism $C' \to Y$ factors through the equalizer of f and g, whence the equalizer is not a proper subobject; $f = g$. Also $\underline{D}$ is clearly left wide-complete. By 3.3, D is cocompact; right wide-complete; total, and a normal extension. By 3.4, it is maximal normal.

The results known for maximal normal categories with (only) a small separating subcategory are much more rudimentary than for the separating-coseparating situation. However, some more can be mentioned. (1) The converse of 3.5.b. In fact, any legitimate category on which every small-valued colimit functor is representable is maximal normal. This sharpens

3.4 but is no harder to prove. (Imitate the first half of the proof of 3.4, not the second.) (2) Negatively, no left completeness property can be established. For example, dually, the category of complete Boolean algebras has a small coseparating subcategory and is maximal normal, but not [19] right small-complete. The maximal normality is easily proved by means of (1) and illegitimate small-complete Boolean algebras. The same proof applies to any of Linton's *equational categories* [16].

I do not know if every maximal normal extension $\underline{E}$ of a category with a small separating subcategory is a minimal right wide-complete extension. If $\underline{E}$ should be left wide-complete, it is minimally so; for $\underline{E}$ and any left wide-complete subextension $\underline{D}$ have a small separating subcategory (3.7), and $\underline{D}$ is cocompact, maximal normal, and total. Over a category $\underline{C}$ with small separating and coseparating subcategories, this gives the analogue of 1.1: maximal normal extensions are left wide-complete, and minimally so.

Let us conclude this section by stating two theorems involving chain conditions, and proving one. A category is said to satisfy the *left wide ascending chain condition* if every well-ordered ascending chain of subobjects of an object is small. The *left wide descending chain condition* is defined similarly; the corresponding *right* conditions are defined dually.

3.10 Theorem. *For a maximal normal category with small separating and coseparating subcategories, the following conditions are equivalent.*

(a) *Every small-valued grounding preserving small limits is representable.*

(b) *The right wide ascending chain condition.*

(c) The left wide descending chain condition.

Presence of a small generating or cogenerating subcategory implies these conditions and their duals.

This is proved in [12]. The concluding sentence follows from a lemma already cited from [10]; a small generating subcategory (in this setting) implies well-poweredness, hence (c) and the dual of (b).

For the application to completions, there seems to be no analogue of the sharp results 3.5.c, 3.5.d. (To prove that there is no analogue one would need a category some of whose maximal normal extensions, but not all, satisfy the conditions of 3.10. All I mean is that no analogue is apparent.) Clearly if $\underline{C}$ is the right universal closure of a small subcategory and is separated by a small subcategory, then the same holds for each of its maximal normal extensions. Then, because of 3.5.a and the manner of forming wide-closures [10], any maximal normal extension is a minimal left small-complete extension and a minimal right small-complete extension.

As noted in the introduction, the category of CW-complexes satisfies this condition. It is the right universal closure of a 1-simplex, and a 1-simplex separates.

3.11 Theorem. A complete, left wide-complete legitimate category satisfying the left wide ascending chain condition can be retracted upon any cocompact full subcategory by a coreflector followed by a reflector.

Proof. A cocompact full subcategory $\underline{B}$ is in any case a reflection of its full left context $\underline{D}$, since $\underline{B} \to \underline{D}$ preserves limits. So we need a coreflective full subcategory of the given $\underline{E}$ between $\underline{B}$ and D. Let $\underline{C}$ consist of the objects C such that $\underline{B}(\ ,C)$ factors through no proper subobject of C. Since $\underline{E}$ is left wide-complete, it can be coreflected upon $\underline{C}$ by simply taking for each object E its smallest subobject containing $\underline{B}(,E)$. The ascending chain condition secures $\underline{C} \subset \underline{D}$, for the subobjects of any C in $\underline{C}$ which are multistrict quotients of objects of $\underline{B}$ are in $\underline{D}$, are closed under small suprema, since $\underline{D}$ is closed under small coproducts, and are not in a proper ideal of the lattice of subobjects.

Reflections and coreflections preserve some compactness, but not through a composition, in general. Incidentally, though this result could hardly be permitted to settle the terminology, it is conveniently stated the way we have defined compactness. The proof is a simple exercise.

3.12. A full reflection of a compact category is compact.

But reflection does not preserve cocompactness. If it were so, it would follow from the duals of 3.11 and 3.12 that every compact category having a good extension $\underline{E}$ (e.g. compact, cocompact, right wide a.c.c.) is cocompact. The category of groups refutes this. We noted that it is compact and not cocompact. The proof of 2.2.c on the assumption (A) in [11] shows without (A) that a primitive category of algebras can be fully embedded in a complete category having small generating and separating subcategories, which is "good" (as above) by 3.10.

4. The lattice of completions

First, Co(C,S) is somewhat complete.

4.1. *Co(C,S) is wide-complete for all C. If C is small, Co(C,S) is legitimate, well-powered, and extraordinary-injective.*

Proof. For a product P of couples X_α, 'P is the product functor of the 'X_α and P' a disjoint sum of the X_α', with the obvious coupling. Equalizers are constructed in the same rigidly formal way. Left wide-completeness becomes obvious when one checks that T cannot be a monomorphism unless both 'T is monomorphic (one-to-one) and T' epimorphic (onto). The dual is a corollary. If C is small, Co(C,S) is clearly legitimate and well-powered, and any legitimate extension E can be retracted by double subregular representation over C.

4.1.a. For all C, any legitimate retract of Co(C,S) containing C is extraordinary-injective, by trivial modification of the previous proof.

It is unknown whether Co(C,S) must be compact for small C. Note, we have no compactness (Special Adjoint Functor) theorem without at least a small separating or coseparating subcategory (though 3.12 can reach further).

Two subcategories of Co(C,S) containing all normal extensions have, sometimes, better properties than the whole. The first of these (4.2.a) is needed for the complete-lattice theorem 4.5; the second seems likely to help in a variety of special cases, but in the special case to be studied below - - completions of the category of finite sets - - a special argument gives a much stronger result, 4.6.

4.2.a. A normal extension consists of separated couples. (Obvious.)

All separated couples will form a legitimate category CoS(C,S) if C has a small separating or coseparating subcategory, by 3.6 and 3.7. Incidentally, CoS(C,S) is injective whenever it is legitimate, by 4.1. a. So if there are both separating and coseparating small subcategories, CoS(C,S) is compact.

4.2.b. The one-sided subregular representations of a normal extension of C go into the "regular completions". For small C the quotation marks are not needed. In general, a left regular transcompletion and left regular left completion were defined in [10], as well as a tour de force "left regular completion" which need not be legitimate but is a minimal complete extension. Let us not try to fit these into the normal extension theory for arbitrary C; in the present set theory, the transcompletion doesn't exist. In the set theory of [10], it is a routine induction (like 2.3) to prove that left subregular representation takes a normal extension E of C into the left regular transcompletion. Interpreting and returning to the present set theory: one should represent by means of a category Sets big enough to provide two successive conjugates for each small-valued grounding or cogrounding of C; then E is represented subregularly into the full left universal closure of the regular representation of C.

The set functors in the regular transcompletions are called stable. Thus 4.2.b says that normal extensions of any legitimate C in Co(C,S) lie in the couples of stable functors. To exploit it:

4.2.c. Assume again that C has small separating and coseparating subcategories. Then for one thing, the category of grounding couples of stable functors is legitimate, and Sets in 4.2.b can be replaced by S.

(In fact, the category of grounding couples of separated functors is legitimate in this setting. These assertions are exercises in the use of 3.8.) The ordinary stable functors form maximal normal extensions (by 3.5.a) $\underline{E}_0$, $\underline{E}_1$. Since conjugate functors are limit resp. colimit functors (2.1 of [10]), $K_1{}^*$ factors through $\underline{E}_1$, K_2 through $\underline{E}_0$. Restricting the conjugations to $\underline{E}_0$, $\underline{E}_1$, respectively, we get adjoint functors $J_1{}^*: \underline{E}_0 \to \underline{E}_1$, $J_2: \underline{E}_1 \to \underline{E}_0$. (For cutting down domain and codomain to full subcategories preserves adjointness.) Then 4.2.b can be restated (on the present assumption): normal extensions lie in $\underline{Cyl}(J_1{}^*, J_2)$.

Knowing nothing about $\underline{Cyl}$ beyond 2.7, we have no application of 4.2.c beyond the conjunction: all normal extensions are boundable if $\underline{E}_0$ and $\underline{E}_1$ are (and $\underline{C}$ has the small subcategories required for 4.2.c; but that can be reduced to the small subcategory required for 2.7.)

One can improve 4.2.b to prove a "joint stability" for couples in normal extensions (proof rather like 2.3). I omit a precise statement, not knowing how to use it. The best possible result of this sort would be a characterization of the full union of the (maximal) normal extensions in $\underline{Co}(\underline{C},\underline{S})$. The work over finite sets after 4.7 below shows that we are far from such a result.

The intersection of all maximal normal extensions, up to equivalence, was determined over small $\underline{C}$ in [11]. It is the intersection of the two regular completions and consists of the <u>reflexive couples</u>, i.e. pairs of mutually conjugate functors coupled by evaluation.

4.3. <u>For a legitimate category $\underline{C}$ with small separating and coseparating subcategories, no non-reflexive couple is isomorphic with an object of every maximal normal extension in $\underline{Co}(\underline{C},\underline{S})$. If $\underline{C}$ merely has</u>

<u>a small separating subcategory, then a reflexive couple is isomorphic with an object of every maximal normal extension in</u> $\underline{\text{Co}}(\underline{C},\underline{S})$.

This proposition can surprise no reader. Why such strong assumptions? For the first part, the trouble is that we have no general method of constructing maximal normal extensions. (In 3.2, we needed 2.3 to rule out non-terminating growth.) The trouble would vanish if it happened that every extension which is maximal normal had a maximal normal sub-extension; but it does not [12]. Still one may hope to improve this part of 4.3. On the second part, a counterexample in [12] leaves little room for improvement; it is injective, well-powered and co-well-powered.

To prove 4.3, first part, note that a couple occurring both in $\underline{E}_0$ and in $\underline{E}_1$ must be reflexive. For the second part, adjoining the reflexive couple X to a supposed maximal normal extension $\underline{F}$ of $\underline{C}$ in $\underline{\text{Co}}(\underline{C},\underline{S})$ yields a Lambek extension, by 2.1 above and 2.1 of [10] (conjugates are limits), and 3.6 makes it legitimate.

We turn from the way the maximal normal extensions "fill space" to the way they fit together. Note that in studying this matter only within $\underline{\text{Co}}(\underline{C},\underline{S})$, we are not imposing an artificial scaffolding. Two full embeddings of one normal extension $\underline{E}$, each agreeing with the double regular representation on $\underline{C}$, are equivalent in $\underline{\text{Co}}(\underline{C},\underline{S})$, i.e. related by endofunctors T, U of the couple category whose composites TU, UT are naturally equivalent to 1. The proof is an exercise with the double Yoneda Lemma.

On this topic we have nothing without small separating and coseparating subcategories. Let us call the equivalence classes of maximal normal extensions, in this setting, <u>compactification types</u>. We define a quasi-ordering of the maximal normal extensions in $\underline{Co}(\underline{C},\underline{S})$: $\underline{D} \le \underline{E}$ provided $\underline{D}$ is contained in the reflectivizer of $\underline{E}$, i.e. $\underline{E}$ is reflective in their full union. This definition works and agrees with the dual definition, in the following sense.

4.4. For the maximal normal extensions of a category $\underline{C}$ with small separating and coseparating subcategories, subregularly embedded in $\underline{Co}(\underline{C},\underline{S})$, $\underline{D}$ and $\underline{E}$ belong to the same compactification type if and only if $\underline{D} \le \underline{E}$ and $\underline{E} \le \underline{D}$. The relation $\le$ is a quasi-ordering. $\underline{D} \le \underline{E}$ if and only if $\underline{E}$ is contained in the coreflectivizer of $\underline{D}$.

Proof. In defining $\underline{D} \le \underline{E}$ we referred to a reflectivizer because we want a reflector; however, as these categories are cocompact, "reflectivizer" = "full left context". Moreover, $\underline{D}$ is in the full left context of $\underline{E}$ in $\underline{Co}(\underline{C},\underline{S})$ if and only if this is true in their full union $\underline{J}$. Then $\underline{E}$-limits are $\underline{J}$-limits. The full right context of $\underline{D}$ in $\underline{J}$ is closed under $\underline{J}$-limits and contains $\underline{C}$; hence it contains $\underline{E}$. This proves the last sentence of 4.4 and makes the first sentence evident. In particular, $\le$ is a reflexive relation. Assume $\underline{D} \le \underline{E} \le \underline{F}$. We have restrictions of reflectors $Q: \underline{D} \to \underline{E}$, $R: \underline{E} \to \underline{F}$, and restrictions of coreflectors $L: \underline{E} \to \underline{D}$, $M: \underline{F} \to \underline{E}$. Then (Q,L) and (R,M) are adjoint pairs. Hence so is (RQ,LM). Now for D in $\underline{D}$ and F in $\underline{F}$, $Hom(RQ(D),F)$ is $Hom(D,F)$ (naturally, by means of the composite of reflection morphisms $D \to Q(D) \to RQ(D)$) at least if D is in $\underline{C}$; since the adjoint functor RQ preserves colimits, this holds for all D, and $\underline{D} \le \underline{F}$.

4.5 Theorem. The compactification types of a legitimate category with small separating and coseparating subcategories form a complete lattice, possibly illegitimate.

Proof. For the infimum of any set of types represented by
$\underline{D}_\alpha \subset \underline{CoS}(\underline{C},\underline{S})$, take the full right wide-closure $\underline{I}$ of $\underline{C}$ in the full
left wide-closure $\underline{R}$ of $\underline{C}$ in the intersection $\underline{M}$ of the reflectivizers
of the $\underline{D}_\alpha$. The common lower bounds of the $\underline{D}_\alpha$ are just the maximal
normal extensions in $\underline{M}$. In view of 4.2.a, all categories mentioned
except $\underline{C}$ are compact and cocompact, and $\underline{I}$ is a normal extension. Any
maximal normal extension $\underline{E}$ of $\underline{C}$ in $\underline{M}$ reflects into $\underline{R}$. As in 4.4,
this implies that $\underline{R}$ is contained in the full right context of $\underline{E}$; so
$\underline{I} \geq \underline{E}$.

This concludes the general results on normal completions. Last, we
shall examine the category of finite sets, proving two restrictive results.

4.6. Every normal completion of the category $\underline{S}_0$ of finite sets is
a full category of topological spaces (embeddable in Boolean spaces).

4.7. If there are no small measurable cardinals, every normal
completion of $\underline{S}_0$ not equivalent to $\underline{S}$ consists of countably compact
spaces.

4.7.a. The full union of the normal completions of $\underline{S}_0$ in $\underline{Co}(\underline{S}_0,\underline{S})$
is not complete.

In the other direction, we find ∞ compactification types, which
illustrates "possibly illegitimate" in 4.5 but does not really illustrate
4.5, for the types we find form a chain.

$\underline{S}_0$ is not merely a conspicuous small category, but an algebraic
theory in (almost) the sense of·Lawvere [15]. It is the vacuous theory,
with no operations and no axioms. Some of the methods will extend to
other theories. We note that 2.6 (with 4.2.c and 4.9) shows that every
normal completion of an algebraic theory consists of algebras with some

"incompressible" structure: incompressible, in that a morphism inducing an isomorphism of algebras must be an isomorphism. In the setting of 4.6 this means a one-to-one continuous function onto is a homeomorphism. The dual of 2.6 adds more technical information, which will be needed for any more than superficial study of the completions of $\underline{S}_0$.

We shall call a legitimate category $\underline{T}$ a theory if all of its objects are small copowers of one object, a total theory if also every small copower exists. A total theory with a distinguished basic object and with distinguished coordinate injections for the copowers constitutes a varietal theory of Linton [16]. Linton extended from the σ-finite case treated by Lawvere [15].

A theory $\underline{V} \subset \underline{C}$ will be called straight in $\underline{C}$ if the embedding is full and preserves coproducts. Evidently every straight theory $\underline{V}$ in a complete category $\underline{C}$ can be extended to a total theory $\underline{T}$ in $\underline{C}$, uniquely (up to equivalence) if we require $\underline{T}$ to be straight and generated from $\underline{V}$ by copower formation. (Existence of $\underline{T}$ follows if $\underline{V}$ is merely full in $\underline{C}$; it is trivial for large $\underline{V}$, and a small $\underline{V}$ lies in a completion in $\underline{C}$ and is straight in that.)

4.8. In a normal extension of a theory $\underline{T}$, every full theory containing $\underline{T}$ is straight.

For a theory is straight in a normal extension $\underline{E}$ of itself, and $\underline{E}$ is a normal extension of any intermediate theory. The non-uniqueness of the full total theory containing $\underline{T}$ is bounded above by the behavior of idempotent morphisms in $\underline{T}$, since an object is a retract of any coproduct of a non-empty family of copies of itself. Thus in the case awaiting us, $\underline{T} = \underline{S}_0$, the containing full total theory is unique.

The next result 4.9 will not be used in proving 4.6 - 4.7 nor even in finding ∞ compactification types. It is put here to encourage the reader through the bare sketch of proof that the types found form a chain. That proof is tedious but not hard, and it uses 4.9.

Categories of models are much the same in the present generality as in [15] and [16]. A model of a theory $\underline{T}$ is a product-preserving functor $\underline{T}^* \to \underline{S}$. Representable functors are free models. Free models are projective, and each one except those represented by 0-th copowers is a generator. The category of models is complete, thus a complete extension of $\underline{T}$. It is the largest projectively generated complete extension, as follows.

4.9. If a right complete category $\underline{C}$ is generated by a straight total theory $\underline{T}$ of projective objects, then $\underline{C}$ is left complete and well-powered, the left subregular representation over $\underline{T}$ embeds $\underline{C}$ fully in the category of models of $\underline{T}$, and the full left closure of $\underline{T}$ is its left regular left completion. The left regular left completion $\underline{L}$ of any theory $\underline{V}$ lies fully in its category of models, but the objects of $\underline{V}$ need not be projective in $\underline{L}$ - - and are not, for the theory of groups.

The following proof sketch gives a complete proof, but I omit to reproduce a fairly long proof from [9], all of which is needed. The left regular left completion of V is its full left closure in $\underline{Cat}(\underline{V}^*,\underline{S})$. The term $\underline{T}$-sesquistrict quotient means a strict quotient coequalizing (simultaneously) some set of pairs of morphisms with domains in $\underline{T}$.

Two preliminary reductions: $\underline{C}$ will certainly be well-powered, since a straight theory generates $\underline{C}$ only if one of its objects generates; and the full embedding in models means just that $\underline{T}$ is left adequate as well as straight. For adequacy, compare the proof of 5.2 in [9]. The first four paragraphs of it do not use the finiteness hypothesis, and the present hypotheses therefore suffice for those paragraphs. They show that each object X of $\underline{C}$ is a $\underline{T}$-sesquistrict quotient of some T in $|\underline{T}|$, by $q: T \to X$. Any natural transformation $\varphi: \underline{T}(\ ,X) \to \underline{T}(\ ,Y)$ will take q to some $r: T \to Y$, which coequalizes all pairs of $\underline{T}$-morphisms into T coequalized by q (by naturality) and hence factors, $r = fq$. Since the objects of $\underline{T}$ are projective, all of their morphisms to X are left multiples of q. Thus φ is $\underline{T}(\ ,f)$, and $\underline{T}$ is left adequate.

The rest of 5.2 [9] applies if we replace the ordinary (finitary) algebras with infinitary algebras after Linton [16]. Then $\underline{C}$ is closed under formation of small products and arbitrary subalgebras, and therefore is left complete. From this, by 4.4 of [10], $\underline{C}$ is total in its full left closure in $\underline{\text{Cat}}(\underline{T}^*,\underline{S})$; so the left closure of $\underline{T}$ in $\underline{C}$ is equivalent to its left regular left completion.

The affirmative part of the second sentence of 4.9 follows from what was just proved, applied to the category of models of $\underline{V}$ and the straight total extension $\underline{T}$ generated by $\underline{V}$. For the negative, it will suffice to exhibit a subdirect product G of finitely generated free groups not in the left closure of those groups (in the category of models, i.e. of groups). For G will be, like every group, a coequalizer of morphisms $f, g: A \to B$ of free groups. In the left closure in question (right complete by 1.2 or the original Special Adjoint Functor Theorem), f and g have a different coequalizer H. This gives a monomorphism $G \subset H$ that is not an isomorphism. Then a morphism from the group of integers Z to H taking the integer 1 to an element of H that is not in G cannot be lifted up the coequalizer $B \to H$.

Recall that abelian subgroups of free groups are cyclic (indeed, all subgroups are free). Recall (<u>slenderness</u> [2]) that the homomorphisms from a countable direct power P of Z to Z are just the finite sums of coordinate projections. Hence the same holds for the subgroup G of a direct product H of $\aleph_1$ copies of Z which consists of the elements having only countably many non-zero coordinates. Now every homomorphism of G into a free group factors across Z; homomorphisms to Z extend uniquely over H; H is the reflection of G in the left closure of the finitely generated free groups. 4.9 is proved.

By 4.9, free groups (and by its proof, free abelian groups) are not projective in any normal completion of the theory. The complete categories for which they are projective generators are precisely the full categories of groups containing free groups and closed under product and subgroup; this is the gist of 5.2 of [9], which we have just refined and generalized.

Proof of 4.6. For a normal completion $\underline{E}$ of $\underline{S}_0$, we have a double subregular representation X = ('X,X',m). The underlying set functor of the desired topological representation is defined from that and a one-point set $P \in |\underline{S}_0|$, X going to 'X(P); since 'X (contra-) preserves coproducts, this is faithful. Then X' gives a set of finite-valued functions separating points on 'X(P), and this determines a topology and a Boolean compactification. Since morphisms are conjoint transformations, this representation takes them to continuous functions. It remains to prove fullness.

Now we may write X for 'X(P). We identify small cardinals D with initial ordinals; note that this makes D a set of power D. It is convenient to use the same symbol D for a discrete space of that size (which need not be in $\underline{E}$). The coproduct [DP] of D copies of p in $\underline{E}$ contains the set D of coordinate injections (in '[DP](P)); [DP]' is the coproduct functor, and '[DP](P) $\subset$ [DP]'*(P) = βD, the Stone-Čech compactification of D. Since [DP]' is the coproduct functor, the inclusions are topological. Every continuous function from [DP] to a finite space C extends continuously over βD and is therefore a morphism in [DP]'(C) (by 2.3). Next, the set of objects of $\underline{E}$ for which every continuous function to a finite space is a morphism is closed under passage to strict quotients e: X $\to$ Y. For given a continuous function f: Y $\to$ C, fe is continuous, hence a morphism, and fe co-equalizes whatever e does. It follows that fe = ge for some morphism g: Y $\to$ C. Thus f = g on e(X). e(X) is dense, since if it missed an open set it would miss a basic open set, and one could construct morphisms in some Y'(C) to refute the assumption that e is epimorphic. So f = g, f is a morphism. Since a normal completion is a minimal right small-complete extension, all objects have the property in question. Hence every continuous function induces a conjoint transformation and is a morphism.

Proof of 4.7. In a completion $\underline{E}$, consider the free space on $\aleph_0$ points. If it is a discrete space N then every normal completion of the full subcategory $\underline{T}$ on $\underline{S}_0$ and N contains the category of reflexive couples over $\underline{T}$, which is $\underline{S}$, modulo measurable cardinals [7]. Since $\underline{S}$ is complete, $\underline{E}$ is equivalent to $\underline{S}$ in this case. If the free space [NP] properly contains N, then for every sequence of points x_j in a space X of $\underline{E}$, there exist a morphism [NP] $\to$ X taking N to $\{x_j\}$ and a cluster point of $\{x_j\}$.

Proof of 4.7.a. The space of irrational numbers $\underline{T}$ is not in any normal completion of $\underset{\rightharpoondown}{S}_0$; for it has N as a retract, and the discrete space J of the same power as I is reflexive over $\underset{\rightharpoondown}{T}$ of the preceding proof [7], and a one-to-one correspondence $J \to I$ would violate 2.6. Hence there is no product of $\aleph_0$ copies of N in the full union of the normal completions; for the product would map monomorphically to the topological product I, and every convergent sequence in I would factor through it, and convergent sequences determine the topology of I.

To exhibit ∞ compactification types, consider the following categories, for each small regular aleph λ: the category of all topological spaces X embeddable in a Boolean space, such that for every discrete D smaller than λ, every function $f: D \to X$ has a continuous extension $f^{-}: \beta D \to X$, and every finite-valued function $g: X \to C$ such that all gf^{-} are continuous is continuous. It is easy to show that these categories $\underset{\rightharpoondown}{E}_\lambda$ are left wide-complete; so they contain normal completions $\underset{\rightharpoondown}{D}_\lambda$. $\underset{\rightharpoondown}{E}_\lambda$, hence $\underset{\rightharpoondown}{D}_\lambda$, contains just one total theory extending $\underset{\rightharpoondown}{S}_0$, the free space on a generating set D consisting of all points of βD that are limits of less than λ points of D. (The free space must be between D and βD by the proof of 4.6, and this is the only such space in $\underset{\rightharpoondown}{E}_\lambda$.) Thus all $\underset{\rightharpoondown}{D}_\lambda$ are inequivalent.

One can show easily that in $\underset{\rightharpoondown}{E}_\lambda$, strict epimorphisms are onto. Hence free objects are projective, and by 4.9, $\underset{\rightharpoondown}{E}_\lambda$ lies in the category of models of its theory. By 4.4 of [10], $\underset{\rightharpoondown}{D}_\lambda$ can only be the left regular (left) completion of the theory. Then it is easy to show that the $\underset{\rightharpoondown}{D}_\lambda$ form a chain in the lattice of compactification types of $\underset{\rightharpoondown}{S}_0$.

The categories $\underset{\rightharpoondown}{D}_\lambda$, besides consisting entirely of countably compact spaces, contain very many compact spaces. On a rather too strong assumption about cardinals, one has the following.

4.10. _If ∞ is the first uncountable weakly inaccessible cardinal, then the total theory consisting of all free Boolean algebras has ∞ compactification types._

Proof. On this hypothesis, the associated Boolean spaces, direct powers of a two-point space, are already in $\underline{E}_{\aleph_0}$. For every real-valued function on such a product space which is continuous on convergent sequences is continuous [17]. These spaces are evidently limits of finite spaces and strict quotients of free spaces in any $\underline{E}_\lambda$; hence they are in $\underline{D}_\lambda$, and every $\underline{D}_\lambda$ is a compactification of this category.

REFERENCES

[1] P. J. Freyd, Abelian Categories, Harper and Row, New York 1964.

[2] L. Fuchs, Abelian Groups, Akademiai Kiado, Budapest 1958.

[3] D. B. Fuks, Natural mappings of functors in the category of topological spaces (Russian), Mat. Sb. 62 (104) (1963), 160-179.

[4] A. Grothendieck, Sur quelques points d'algèbre homologique, Tôhoku Math. J. (2) 9 (1957), 119-221.

[5] M. Hasse and L. Michler, Theorie der Kategorien, VEB, Berlin 1966.

[6] Z. Hedrlin and A. Pultr, On full embeddings of categories of algebras, Illinois J. Math. 10 (1966), 392-406.

[7] J. Isbell, Adequate subcategories, Illinois J. Math. 4 (1960), 541-552.

[8] ...., Two set-theoretical theorems in categories, Fund. Math. 52 (1963), 43-49.

[9] ...., Subobjects, adequacy, completeness and categories of algebras, Rozprawy Mat. 36 (1964), 1-32.

[10] ...., Structure of categories, Bull. Amer. Math. Soc. 72 (1966), 619-655.

[11] ...., Small adequate subcategories, J. London Math. Soc., to appear.

[12] ...., Small subcategories and completeness, submitted to Math. Systems Theory.

[13] D. Kan, Adjoint functors, Trans. Amer. Math. Soc. 87 (1958), 294-329.

[14] J. Lambek, Completions of Categories, Lecture Notes in Mathematics 24 (Springer), 1966.

[15] F. W. Lawvere, Functorial semantics of algebraic theories, thesis, Columbia Univ., 1963; summarized in Proc. Nat. Acad. Sci. U.S.A. 50 (1963), 869-872.

[16] F. E. J. Linton, Some aspects of equational categories, Proc. of the Conference on Categorical Algebra (La Jolla 1965), New York 1966, pp. 84-94.

[17] S. Mazur, On continuous mappings on Cartesian products, Fund. Math. 39 (1952), 229-238.

[18] Z. Semadeni, Projectivity, injectivity, and duality, Rozprawy Mat. 35 (1963), 1-47.

[19] R. Solovay, New proof of a theorem of Gaifman and Hales, Bull. Amer. Math. Soc. 72 (1966), 282-284.

Case Western Reserve University, Cleveland, Ohio

/bar
22 June 1967

LOCALLY DISTRIBUTIVE SPECTRAL CATEGORIES AND STRONGLY REGULAR RINGS

(Preliminary report.)

by

Jan-Erik ROOS

INTRODUCTION.- Let $\underline{C}$ be a Grothendieck category in the terminology of [12], i.e. $\underline{C}$ is an abelian category which has a family of generators as well as (filtered) direct limits that are exact. Such a category has sufficiently many injectives [15], and it is a rather natural problem to determine the behaviour of these injectives with respect to decomposition, passage to inverse and direct limit etc.

Let $\text{Spec}(\underline{C})$ be the category obtained from $\underline{C}$ by making formally invertible all essential monomorphisms in $\underline{C}$, and let $\underline{C} \xrightarrow{P} \text{Spec}(\underline{C})$ be the natural functor [cf. § 1 below]. This category $\text{Spec}(\underline{C})$ [the spectral category of $\underline{C}$] was introduced in [12], and the properties of injectives in $\underline{C}$ can be conveniently expressed in terms of the pair $(P, \text{Spec}(\underline{C}))$. The structure of $\text{Spec}(\underline{C})$ was essentially determined in [35], [36], and the aim of this paper is to provide proofs of the principal results of [35] and to develop [35] into a still more explicit theory and to give more examples and applications. There are still many questions unanswered and this paper must therefore be considered as a preliminary step to a longer one that would also develop [36].

In order to make this report reasonably self-contained, we will begin by a quick review of [12].

§ 1. - Review and completion of known results and definitions about spectral categories and regular rings.

Let $\underline{C}$ be a Grothendieck category. Recall that a monomorphism $C_1 \to C_2$ in $\underline{C}$ is said to be an essential monomorphism, if for every non-zero subobject C of C_2, $C \times_{C_2} C_1$ is a non-zero subobject of C_1. It is known [11] that every object C of $\underline{C}$ admits an essential monomorphism into an (essentially unique) injective object (the injective envelope of

C). Let $\underline{C} \xrightarrow{P} \operatorname{Spec}(\underline{C})$ be the solution of the universal problem of making all essential monomorphisms in $\underline{C}$ invertible [12]. It is proved in [12] that we can take

$$\operatorname{Ob}(\operatorname{Spec}(\underline{C})) = \operatorname{Ob}(\underline{C})$$

and the functor P to be the identity on objects, and

$$\operatorname{Hom}_{\operatorname{Spec}(\underline{C})}(P(C_1), P(C_2)) = \varinjlim_{C_1' \hookrightarrow C_1} \operatorname{Hom}_{\underline{C}}(C_1', C_2)$$

where we take the direct limit over all essential monomorphisms $C_1' \hookrightarrow C_2$, and P on morphisms is then defined in the natural way. The category $\operatorname{Spec}(\underline{C})$ turns out to be a Grothendieck category of cohomological dimension zero (called hereafter a spectral category), and every such Grothendieck category can of course be obtained in this way. Further the functor P is left exact and it commutes with arbitrary directed unions (in particular with arbitrary direct sums). However P is not exact unless it is an equivalence of categories, and in fact it is easy to see that the cohomological dimension of $\underline{C}$ is equal to the least n such that $R^{n+1}P = 0$. We may also remark that one verifies without difficulty that a short exact sequence

(1) $$0 \longrightarrow C_1 \longrightarrow C_2 \longrightarrow C_3 \longrightarrow 0$$

is transformed into an exact sequence by P, if and only if the sequence (1) is high in the terminology of [41] or if and only if C_1 is a complementary submodule of C_2 in the sense of [31]. This observation can be used to simplify some of the proofs of [41] and [31]. For example, the fact that the high sequences form a proper class [41] follows now directly from the general theory of relative homological algebra (see e.g. [9] and [20]).

It is readily verified that two objects C_1 and C_2 in $\underline{C}$ have isomorphic injective envelopes if and only if $P(C_1)$ is isomorphic to $P(C_2)$ [12], and that the injective envelope of C contains as an essential subobject [1] a sum of indecomposable injectives if and only if P(C) is a direct sum of simple object in Spec($\underline{C}$) [22]. Furthermore Spec($\underline{C}$) decomposes in an essentially unique way into a product of spectral categories

$$\text{Spec}(\underline{C})_{dis} \times \text{Spec}(\underline{C})_{cont}$$

where $\text{Spec}(\underline{C})_{dis}$ [the discrete part of Spec($\underline{C}$)] is equivalent to a product category $\prod_{\alpha \in I} \text{Mod}(K_\alpha)$ [2] where the K_α:s are skewfields, and where $\text{Spec}(\underline{C})_{cont}$ [the continuous part of Spec($\underline{C}$)] does not contain any simple object [12].

It should be remarked that P does not commute with arbitary infinite products or with arbitary (filtered) $\varinjlim$ even if $\text{Spec}(\underline{C})_{cont} = 0$, in fact one can even prove the following more precise results for a general Grothendieck category $\underline{C}$:

(A) $\underline{C}$ is locally noetherian in the sense of [11] i.e. $\underline{C}$ has a family of noetherian generators if and only if

(i) $\text{Spec}(\underline{C})_{cont} = 0$;

(ii) P commutes with arbitrary filtered $\varinjlim$.

(B) $\underline{C}$ is locally finite in the sense of [11] i.e. $\underline{C}$ has a family of generators of finite length if and only if $\underline{C}$ satisfies (i) and (ii) above as well as

(iii) P commutes with arbitrary $\varprojlim$ (or, which is equivalent, P commutes with arbitary products).

(C) $\underline{C}$ has Krull dimension zero in the sense of [11] if and only if $\underline{C}$ satisfies (i) and (iii) above.

(See also the Remark 3 at the end of § 5 below.)

In [12] the structure of spectral categories was intimately related to the rings that are regular in the sense of von Neumann. Recall that a

---

1) We say that the subobject D of C is an essential subobject if the natural monomorphism $D \to C$ is an essential monomorphism.

2) If Λ is a unitary ring then Mod(Λ) will always denote the category of right unitary Λ-modules.

ring R is said to be regular in the von Neumann sense if for every $a \in R$ there is an $x \in R$ such that $a = axa$. It is equivalent to say that the weak homological dimension of R is zero [3] or that every principal right (left) ideal of R is a direct factor, i.e. is generated by an idempotent [23]. In [12] it was proved that the endomorphism ring of an arbitrary object in a spectral category $\underline{D}$ is a regular right self-injective ring. Furthermore, if G is a generator of $\underline{D}$, then the localizing subcategory [11] $\underline{L}$ of $\mathrm{Mod}(\mathrm{Hom}_{\underline{D}}(G,G))$ in the Gabriel-Popescu representation [28], [12], [33] of $\underline{D}$:

$$0 \longrightarrow \underline{L} \longrightarrow \mathrm{Mod}(\mathrm{Hom}_{\underline{D}}(G,G)) \longrightarrow \underline{D} \longrightarrow 0$$

is defined by the family of essential right ideals of the regular right self-injective ring $\mathrm{Hom}_{\underline{D}}(G,G)$.

Conversely, given any regular right self-injective ring R, then the essential right ideals of R define a localizing subcategory $\underline{L}$ of Mod(R), and so we have as before an "exact sequence" [33]

$$0 \longrightarrow \underline{L} \longrightarrow \mathrm{Mod}(R) \underset{j_*}{\overset{j^*}{\rightleftarrows}} \mathrm{Mod}(R)/\underline{L} \longrightarrow 0$$

where j_* is a right adjoint to j^*. Here $\mathrm{Mod}(R)/\underline{L}$ is a spectral category and more precisely j_* defines an equivalence between this category and the full subcategory of Mod(R) whose objects are the direct factors of products $\prod_I R$. Finally the endomorphism ring of the generator $j^*(R)$ of $\mathrm{Mod}(R)/\underline{L}$ is isomorphic to R.

In what follows we will also need the <u>strongly regular rings</u> that were introduced in [1]. Recall that a ring R is said to be strongly regular if for any $a \in R$ there is an $x \in R$ such that $a = a^2x$. A commutative regular ring, in particular a Boolean ring is strongly regular, as well as an arbitrary product of skewfields. The general case is essentially a mixture of these two later cases, were we allow a "twisting" of the skewfields. For a more precise statement, see § 3. The following proposition is essentially well-known [cf. p. 463 of [1] and [23]].

PROPOSITION 1. - The following conditions on a ring R are equivalent:

(i) R is strongly regular;

(ii) [resp. (ii)'] Every principal right (resp. left) ideal of R is generated by an idempotent that belongs to the center of R.

COROLLARY. - If R is a strongly regular ring, then the lattice of principal right (left) ideals of R is a Boolean algebra and it is even a complete Boolean algebra if R is right self-injective.

Remark 1. - The last property on the Boolean algebra does not imply that R is right self-injective (see § 3 and [27]).

Remark 2. - In § 3 we will see that a strongly regular ring is right self-injective if and only if it is left self-injective.

Remark 3. - The strongly regular self-injective rings are related to a special case of spectral categories that we will study in § 2. This study will also give an extension of Proposition 1 above.

DEFINITION 1. - A regular ring R is said to be locally strongly regular to the right if every non-zero principal right ideal contains a non-zero idempotent e such that eRe is strongly regular.

Finally we recall that if $\overline{R}$ is the injective envelope of a regular ring R considered as a module over itself to the right, then $\overline{R}$ has a natural ring structure for which $R \to \overline{R}$ is a ring homomorphism and for which $\overline{R}$ becomes a regular right self-injective ring [11], that is even strongly regular (resp. locally strongly regular) if R is so. However, $\overline{R}$ is not in general left self-injective, even in the locally strongly regular case (see § 3 below).

§ 2. - Distributive objects in spectral categories.

DEFINITION 2. - Let C be an object in a Grothendieck category $\underline{C}$. We say that C is a distributive object in $\underline{C}$, if the lattice of subobjects of C is a distributive lattice. We say that $\underline{C}$ is locally distributive if $\underline{C}$ has a family of generators consisting of distributive objects.

Example: The commutative rings R that are distributive in Mod(R) have been determined in [18]. The situation is not so simple in the non-commutative case [39].

Now if $\underline{C}$ is a spectral category, then the condition of distributivity can be reformulated in several ways:

PROPOSITION 2. - <u>The following conditions on an object</u> C <u>in a spectral category</u> $\underline{C}$ <u>are equivalent</u>:

(i) C <u>is distributive in</u> $\underline{C}$;

(i)' <u>For every</u> $C_3 \subset C$ <u>and every decomposition</u> $C = C_1 \amalg C_2$, <u>the natural map</u>

(*) $$(C_3 \cap C_1) \amalg (C_3 \cap C_2) \longrightarrow C_3$$

<u>is an isomorphism</u>;

(ii) <u>For any decomposition</u> $C = C_1 \amalg C_2$ <u>we have</u> $\mathrm{Hom}_{\underline{C}}(C_1, C_2) = 0$;

(ii)' <u>If</u> $V_1 \subset C$, $V_2 \subset C$ <u>and</u> $V_1 \cap V_2 = 0$, <u>then</u> $\mathrm{Hom}_{\underline{C}}(V_1, V_2) = 0$;

(iii) <u>The endomorphism ring of</u> C <u>is a strongly regular ring.</u> [Then the ring is also right and left self-injective (cf. § 1 and § 3).];

(iv) [resp. (iv)'] <u>The lattice of subobjects of</u> C <u>is a Boolean algebra</u> (resp. a complete Boolean algebra).

<u>Remark.</u> - In any spectral category $\underline{C}$, $\mathrm{Hom}_{\underline{C}}(C_1, C_2) = 0 \Leftrightarrow \mathrm{Hom}_{\underline{C}}(C_2, C_1) = 0$.

PROOF OF PROPOSITION 2: It is clear that (i) => (i)'. To prove that (i)' => (ii), let $C_3 \subset C_1 \prod C_2 = C_1 \amalg C_2$ be the graph of a morphism $f: C_1 \to C_2$. The formula (*) then implies that this is the graph of the zero morphism, and so (i)' => (ii). To prove that (ii) => (iii), we have by Proposition 1 to prove that every idempotent of $\mathrm{Hom}_{\underline{C}}(C, C)$ is a central element. Thus let $e \in \mathrm{Hom}_{\underline{C}}(C, C)$ be an idempotent, and $C = e(C) \amalg (1-e)(C)$ the corresponding decomposition of C. Now let $f : C \to C$ be an arbitrary element of $\mathrm{Hom}_{\underline{C}}(C, C)$. Then by (ii) the projection on $(1-e)(C)$ of the restriction of f to $e(C)$ must be the zero map, and so $(1-e)fe = 0$. In the same way we get $ef(1-e) = 0$, so that $fe = ef$ and e is in the center and so (ii) => (iii). The

Corollary to Proposition 1 now gives the implication (iii) => (iv)', and since (iv)' => (iv) => (i) and (ii) <=> (ii)' are trivial, the proposition follows.

COROLLARY. - *Let* C_1 *and* C_2 *be two distributive objects in the spectral category* $\underline{C}$. *Then* $C_1 \amalg C_2$ *is distributive if and only if* C_1 *and* C_2 *are, and* $\mathrm{Hom}_{\underline{C}}(C_1,C_2) = 0$.

The corollaries of the following proposition give another way of building up big distributive objects in a spectral category. Recall that a direct system $\{C_\alpha\}$ is said to be a *mono*-direct system [34] if for $\alpha < \beta$ the morphisms $C_\alpha \to C_\beta$ are monomorphisms.

PROPOSITION 3. - *Let* $\{D_\alpha\}$ *be a filtered mono-direct system of distributive objects in the spectral category* $\underline{C}$, *and let* V_1 *and* V_2 *be two subobjects of* $\varinjlim D_\alpha$. *Consider the* D_α:s *as subobjects of* $\varinjlim D_\alpha$. *Then every map* $V_1 \cap D_\alpha \xrightarrow{f} V_2$ *factors in a (necessarily) unique way through the monomorphism* $V_2 \cap D_\alpha \to V_2$, *and so in particular we have a natural map*

$$\text{(2)} \qquad \mathrm{Hom}_{\underline{C}}(V_1,V_2) \to \varprojlim_{\alpha} \mathrm{Hom}_{\underline{C}}(V_1 \cap D_\alpha,\ V_2 \cap D_\alpha)$$

that is an isomorphism.

PROOF: We have $V_2 = \bigcup_\beta (V_2 \cap D_\beta)$ by the axiom AB 5. Consider the pull-back diagrams

$$\text{(3)} \qquad \begin{array}{ccc} V_1 \cap D_\alpha & \xrightarrow{f} & V_2 \\ \uparrow & & \uparrow \\ (V_1 \cap D_\alpha) \times_{V_2} (V_2 \cap D_\beta) & \xrightarrow{f_\beta} & V_2 \cap D_\beta \end{array}$$

We obtain a direct system of morphisms $\{f_\beta\}$, and $\varinjlim f_\beta = f$ by the axiom AB 5. I claim that for $\beta \geq \alpha$, f_β factors in a (necessarily) unique way through the monomorphism $V_1 \cap D_\alpha \to V_2 \cap D_\beta$. In fact we have $D_\beta = D_\alpha \amalg D'_\alpha$ for some subobject D'_α of D_β, and since D_β is distributive we get

$$\text{(4)} \qquad V_2 \cap D_\beta = (V_2 \cap D_\alpha) \amalg (V_2 \cap D'_\alpha) \quad \text{and} \quad \mathrm{Hom}_{\underline{C}}(D_\alpha,\ D'_\alpha) = 0.$$

But the last part of (4) implies in particular that

$$\text{(5)} \qquad \mathrm{Hom}_{\underline{C}}((V_1 \cap D_\alpha) \times_{V_2} (V_2 \cap D_\beta),\ V_2 \cap D'_\alpha) = 0$$

for $V_2 \cap D'_\alpha$ is a subobject of D'_α and by (3) $(V_1 \cap D_\alpha) \times_{V_2} (V_2 \cap D_\beta)$ is a subobject of $V_1 \cap D_\alpha$, thus of D_α. Now it is clear that (5) implies in particular that the projection of f_β on $V_2 \cap D'_\alpha$ is zero, and so by the first part of (4) f_β factors through $V_2 \cap D_\alpha \to V_2 \cap D_\beta$ and the uniqueness of the factorization gives us now a direct system of factorizations for $\beta \geq \alpha$. Now if we pass to the limit over β, we obtain the first part of Proposition 3. The second part follows immediately from this first part, since the AB 5-isomorphism $V_1 = \bigcup_\alpha (V_1 \cap D_\alpha)$ gives rise to an isomorphism

$$\mathrm{Hom}_{\underline{C}}(V_1,V_2) \xrightarrow{\sim} \varprojlim \mathrm{Hom}_{\underline{C}}(V_1 \cap D_\alpha, V_2).$$

COROLLARY 1. - <u>The direct limit of a filtered direct system of distributive objects in a spectral category is again a distributive object.</u>

PROOF: If $\{D_\alpha\}$ is the direct system of distributive objects, and if $E_\alpha = \mathrm{Im}\,(D_\alpha \to \varinjlim D_\alpha)$, then the E_α:s are distributive. Since further these E_α:s form a mono-direct system in a natural way, and since $\varinjlim E_\alpha = \varinjlim D_\alpha$, it is clear that we may suppose that the given direct system is a mono-system. Now let V_1 and V_2 be two subobjects of $\varinjlim D_\alpha$ with $V_1 \cap V_2 = 0$. Since D_α is distributive, and since $(V_1 \cap D_\alpha) \cap (V_2 \cap D_\alpha) = 0$, we get by Proposition 2,(ii)' that $\mathrm{Hom}_{\underline{C}}(V_1 \cap D_\alpha, V_2 \cap D_\alpha) = 0$ for all α, and so $\mathrm{Hom}_{\underline{C}}(V_1,V_2) = 0$ by formula (2) of Proposition 3 and therefore $\varinjlim D_\alpha$ is distributive by Proposition 2,(ii)'.

COROLLARY 2. - <u>Let</u> C <u>be an object in a spectral category and let</u> D <u>be a distributive subobject of</u> C, <u>and consider the set of distributive subobjects of</u> C <u>that contain</u> D, <u>partially ordered by inclusion. This partially ordered set has a maximal element.</u>

PROOF: According to Corollary 1 and the axiom AB 5, we can apply Zorn's lemma, and the result follows.

Now we come to the essential result of this section.

THEOREM 1. - <u>Let</u> $\underline{C}$ <u>be a spectral category. Then</u> $\underline{C}$ <u>admits a</u>

distributive object D_{max} that is maximum in the sense that every distributive object of $\underline{C}$ is isomorphic to a subobject of D_{max}. The object D_{max} is unique up to isomorphism. Let $\underline{C}_{adistr}$ denote the localizing subcategory of $\underline{C}$ formed by the objects C such that $Hom_{\underline{C}}(C,D_{max}) = 0$ [11], and let $\underline{C}_{distr}$ be the smallest localizing subcategory of $\underline{C}$ that contains D_{max} [11]. Then $\underline{C}_{distr}$ is a locally distributive spectral category, and $\underline{C}_{adistr}$ is a spectral category that does not contain any non-zero distributive object (such a spectral category will be called anti-distributive in what follows) and the natural functor [11]

$$\underline{C} \to \underline{C}_{distr} \times \underline{C}_{adistr}$$

is an equivalence of categories.

PROOF: The proof will be based on the preceeding theory and on three lemmas:

LEMMA 1. - Let G be a generator of a spectral category $\underline{C}$. Then every distributive object D of $\underline{C}$ is isomorphic to a subobject of G.

PROOF OF LEMMA 1: Consider the set of pairs (U,f) where U is a subobject of G and f is a monomorphism $U \xrightarrow{f} D$. These pairs form a partially ordered set in a natural way (extensions of mappings). Furthermore this set is inductive as one easily sees using the axiom AB 5, and so it has a maximal element (U_o,f_o) by Zorn's lemma. I claim that $f_o(U_o) = D$. In fact, if this is not the case, then $D = f_o(U_o) \amalg D'$ where $D' \neq 0$. But G is a generator, and so there is a non-zero map $G \to D'$. Write $G = U_o \amalg U'_o$. Now $U_o \xrightarrow{\simeq} f_o(U_o)$ and so the restriction of φ to U_o must be zero, for otherwise there would be a non-zero map $f_o(U_o) \to D'$, which is impossible, since D is distributive. Thus the restriction of φ to U'_o is a non-zero map $\check{\varphi}: U'_o \to D'$. But from this map it is easy to construct a non-zero monomorphism from a subobject of U'_o to D', and the direct sum of this morphism and f_o extends f_o non-trivially, contradicting the maximality of (f_o,U_o). Thus we have

$$\begin{array}{ccc} G & & \\ \uparrow & & \\ U_o & \xrightarrow{f_o} & D \end{array}$$

where f_o is an isomorphism, so that D is isomorphic to a subobject of G, as we asserted.

LEMMA 2. - *Let* C *be an object in a spectral category,* D *a maximal distributive subobject of* C (cf. Corollary 2 above) *and* D_1 *an arbitrary distributive subobject of* C. *Then* D_1 *is isomorphic to a subobject of* D.

PROOF: Consider the set of pairs (V,f) where V is a subobject of D and where f is a monomorphism $V \xrightarrow{f} D_1$, and order this set by extension of mappings. By the axiom AB 5 the order is inductive, and so by the Zorn lemma we have a maximal element $V_o \xrightarrow{f_o} D_1$. I claim that f_o is an epimorphism (which will prove the lemma, since then f_o is an isomorphism). If f_o is not an epimorphism, then $D_1 = \mathrm{Im}\, f_o \amalg D_1'$, where $D_1' \neq 0$. Then we must have $\mathrm{Hom}_{\underline{C}}(D,D_1') = 0$, for if this group were zero, then also $\mathrm{Hom}_{\underline{C}}(D_1',D) = 0$, and so $D_1' \cap D = 0$, and then $D_1' \amalg D$ could be realized as a distributive subobject of C (Corollary to Proposition 2) which contradicts the maximality of D. Thus $\mathrm{Hom}_{\underline{C}}(D,D_1') \neq 0$, and so let φ be a non-zero morphism $D \to D_1'$. The restriction of φ to V_o must be zero, for $V_o \xrightarrow{\sim} \mathrm{Im}\, f_o$ and $\mathrm{Hom}_{\underline{C}}(\mathrm{Im}\, f_o, D_1') = 0$, since D_1 is distributive. Write $D = V_o \amalg D'$. Now it is clear that we can construct a non-zero monomorphism $D' \xrightarrow{g} D_1'$ from the restriction of φ to D' and the direct sum of this morphism g and f_o is strictly bigger than (V_o,f_o) which is impossible, and so $D_1' = 0$, and the Lemma 2 is proved.

LEMMA 3 ("SCHRÖDER-BERNSTEIN"). - *Let* D_1 *and* D_2 *be two distributive objects in a spectral category, and suppose that* D_1 *is isomorphic to a subobject of* D_2 *and that* D_2 *is isomorphic to a subobject of* D_1. *Then* D_1 *and* D_2 *are isomorphic.* (It is sufficient to suppose that one of the D_i:s is distributive, for then the other one will also be so.)

PROOF: If $D_1 \xrightarrow{i_1} D_2$ and $D_2 \xrightarrow{i_2} D_1$ are monomorphisms, but none of them an isomorphism, then we get by composition a monomorphism $D_1 \xrightarrow{i} D_1$ that is not an epimorphism. But since D_1 is distributive, it follows that the restriction of i to the complement of $Im(i)$ in D_1 must be zero, which is a contradiction since i is a monomorphism, and so the Lemma 3 is also proved.

PROOF OF THEOREM 1: Let G be a generator for $\underline{C}$, and D_G a maximal distributive subobject of G (Corollary 2 of Proposition 3). Let D be an arbitrary distributive object of $\underline{C}$. By Lemma 1, D is isomorphic to a subobject of G, and by Lemma 2 this subobject is isomorphic to a subobject of D_G, and so D_G is maximum in the sense described in Theorem 1. That D_G is unique up to isomorphism follows now from Lemma 3. The rest of Theorem 1 is automatic (any localizing subcategory of a spectral category is a direct factor).

COROLLARY 1. - Any locally distributive spectral category is determined up to equivalence by the isomorphy class of the endomorphism ring of a maximal distributive object.

PROOF: In this case the maximal distributive object is a generator and so we can apply § 1.

Since the endomorphism ring of Corollary 1 can be an arbitrary strongly regular self-injective ring (cf. § 1 and Proposition 2) we get

COROLLARY 2. - There is a natural one-one correspondence between the (equivalence classes of) locally distributive spectral categories and the (isomorphy classes of) strongly regular rings that are self-injective.

COROLLARY 3. - Every right self-injective regular ring R admits an essentially unique decomposition into a product of rings $R \xrightarrow{\sim} R_1 \times R_2$, where R_1 is a locally strongly regular right self-injective ring (see § 1) and where R_2 is a regular right self-injective ring that contains no non-zero idempotent e such that eR_2e is strongly regular.

The structure of the part R_1 of R will be determined in the next section, but first we will study a particularly instructive special case of the locally distributive spectral categories.

DEFINITION 3. - An object C in a category $\underline{C}$ is said to be idempotent in $\underline{C}$, if every endomorphism of C is idempotent. A spectral category $\underline{C}$ is said to be locally idempotent if it has a family of generators that are idempotent.

It is clear that in a spectral category an idempotent object is in particular distributive. Furthermore, we can develop the preceeding theory for idempotent objects instead of distributive ones. This gives us a decomposition analogous to that of theorem 1, in particular:

(6)
$$\underline{C}_{dis} \xrightarrow{\sim} \underline{C}_{idem} \times \underline{C}_{dis,n-idem}$$

where $\underline{C}_{dis,n-idem}$ is a locally distributive spectral category that contains no non-zero idempotent objects, and where $\underline{C}_{idem}$ is a locally idempotent spectral category. Such a category has a maximal idempotent generator that is unique up to a unique isomorphism, and its endomorphism ring is a complete Boolean ring. Since a Boolean ring is self-injective if and only if it is complete [6], we get as a special case of Corollary 2 above a one-one correspondence between the (... classes of) complete Boolean algebras and the locally idempotent spectral categories, and so many theorems about Boolean algebras can be obtained as special cases of theorems about general Grothendieck categories. This is in particular the case of the theory of Smith and Tarski [40] and Pierce [25], [26] about the distributivity of $\cap$ with respect to $\cup$ in a Boolean algebra (cf. also [38]), that is a special case of the theory of distributivity of $\varprojlim$ with respect to $\varinjlim$ in Grothendieck categories (cf. [32] for the non-additive case).

If $\underline{C}_B$ is the spectral category corresponding to the complete Boolean ring B, then one can even show that $\underline{C}_B$ is equivalent to the category of $\underset{\sim}{Z}_2$-sheaves on the category $\underline{B}$ associated to the partially ordered set associated to B, the category $\underline{B}$ being equipped with its canonical topology (cf. § 3 and [37]), but this will be a special case of a more general structure theory concerning locally distributive spectral categories, that we will develop in the next section.

Remark. - It is clear that (6) also gives a decomposition of locally strongly regular right self-injective rings, that is analogous to Corollary 3 above.

§ 3. - On the explicit structure of strongly regular rings and their associated spectral categories.

In what follows all rings will have a unit, some of these results extend however to the non-unitary case [8].

The first use of sheaf theory in connection with regular rings in the von Neumann sense seems to stem from Bourbaki ([5], Exercises 16 and 17 on p. 173), who restricted himself to commutative rings. More generally Dauns and Hofmann [8] have proved that if R is a biregular ring (i.e. a ring where every twosided principal ideal is generated by a central idempotent [1]), then R is isomorphic to the ring of global sections of a sheaf of (Jacobson-)simple rings on the maximal ideal space of R, which is a Boolean space, i.e. a compact totally disconnected space. Furthermore the converse of this is true. Later Pierce [27] has studied the commutative case further. If we specialize [8] and generalize [5] and [27] then we obtain:

THEOREM 2. - Consider the category τ of ringed spaces (X,κ), where X is a Boolean space, and where κ is a sheaf of skewfields over X, and let the functor $\tau \xrightarrow{\Gamma}$ (Rings) be defined by $(X,\kappa) \longrightarrow \Gamma(X,\kappa)$ [14]. This functor defines an equivalence between the category τ and the full subcategory of (Rings) formed by the strongly regular rings. A quasi-inverse of this restricted Γ is obtained by associating to the strongly regular ring R the space X = Spec(R) of the twosided maximal (= prime [2]) ideals of R with its usual topology and equipped with the natural sheaf

1) Every strongly regular ring is biregular, but biregular does not imply regular, nor does regular imply biregular.

2) Every left or right ideal of a strongly regular ring is twosided and every quotient with respect to a maximal ideal is a skewfield.

of skewfields whose stalks are R/m, $m \in X$. Furthermore, if (X,κ) and R correspond to each other, then the section functor on modules defines an equivalence between the category $\underline{Mod}_X(\kappa)$ of sheaves of κ-modules over X (every stalk over x is a vector space over $\kappa(x)$, the stalk of κ over x) and the category $Mod(R)$, and in this equivalence the flabby ("flasque" in the terminology of [14]) sheaves (flabby as sheaves of abelian groups) in $\underline{Mod}_X(\kappa)$ correspond exactly to the injective modules of $Mod(R)$.

COROLLARY 1. - A strongly regular ring R is right self-injective if and only if it is left self-injective (and then by the analogue of [27] the associated Boolean space must be complete, i.e. correspond to a complete Boolean algebra in the Stone representation [38]).

In fact, the sheaf κ associated to R is the same as an abelian sheaf in the two cases (left and right), and so the Corollary 1 follows from the last part of Theorem 2.

Now we will use Theorem 2 to describe sheaf-theoretically the spectral category $\underline{C}_R$ associated to a strongly regular self-injective ring R. Recall that by § 1 we have an "exact sequence"

$$(7) \qquad 0 \longrightarrow \underline{L} \longrightarrow Mod(R) \underset{j_*}{\overset{j^*}{\rightleftarrows}} \underline{C}_R \longrightarrow 0$$

where $\underline{L}$ is the localizing subcategory of $Mod(R)$ determined by the essential right ideals of R, and so $\underline{C}_R$ is equivalent by j_* to the full subcategory of $Mod(R)$ whose objects are those M such that the restriction map

$$Hom_R(R,M) \to Hom_R(\mathfrak{a},M)$$

is an isomorphism for all essential right ideals $\mathfrak{a}$ of R.

Every equivalence of categories preserves essential monomorphisms, and since the equivalence $\underline{Mod}_X(\kappa) \xrightarrow{\sim} Mod(R)$ of Theorem 2 transforms κ into R, and since the essential subobjects of κ are easily seen to be the sheaves of the form κ_U [14], [15], where U is open and dense in X, we obtain that $\underline{C}_R$ is equivalent to the full subcategory of $\underline{Mod}_X(\kappa)$

whose objects are those sheaves $\mathcal{F}$ such that the restriction map

(8) $$\Gamma(X, \mathcal{F}) \to \Gamma(U, \mathcal{F})$$

is an isomorphism for all open dense subsets U of X. This can be interpreted in another way:

Let X be any topological space, and consider for any open subset U of X, the set $J(U)$ of families of open subsets of U, $\{U_\alpha \to U\}_{\alpha \in K}$ such that $\overline{\bigcup U_\alpha} = \overline{U}$ ($^-$ denotes closure). It is easy to see that these $J(U)$ (for varying U) define a topology in the sense of Grothendieck [2], [13], [37], [21] on the category Open(X) formed by the open sets of X, and the elements of $J(U)$ will hereafter be called the coverings of U for this topology. One could define the topology and the coverings in more invariant form by means of the notion of "crible" [13], [37], but this is not essential here.

This topology on Open(X) will be called the hypercanonical topology, for it is easy to see that it is the finest [37] topology on Open(X) for which the set-valued presheaves $\mathrm{Hom}_{\mathrm{Open}(X)}(\,\cdot\,, U)$ are separated presheaves (cf. [37] for the terminology), whereas the usual topology (defined by ordinary coverings) is the finest topology for which the $\mathrm{Hom}_{\mathrm{Open}(X)}(\,\cdot\,, U)$ are sheaves [37] and this is generally called the canonical topology on Open(X) [37].

If we now go back to the case where X is a complete Boolean space, it is easy to see that the isomorphism condition on the morphism (8) expressed above is equivalent to the condition that $\mathcal{F}$ is a sheaf for the hypercanonical topology. Thus we can summarize and reformulate our results:

THEOREM 3. - Let X be a complete Boolean space, κ a sheaf of skew-fields over X that is also a sheaf for the hypercanonical topology on X, and let $\mathrm{Mod}_{X,h}(\kappa)$ be the category of sheaves of κ-modules for the hypercanonical topology on X. Then $\mathrm{Mod}_{X,h}(\kappa)$ is a locally distributive spectral category. Conversely, every locally distributive spectral category is obtained in this way from an essentially unique pair (X,κ).

Remark 1. - If R is a strongly regular self-injective ring, then the natural functor $\mathrm{Mod}(R) \xrightarrow{j^*} \underline{C}_R$ can be interpreted as a functor

$$\underline{\mathrm{Mod}}_X(\kappa) \to \underline{\mathrm{Mod}}_{X,h}(\kappa)$$

and this is just the functor that transforms every sheaf of κ-modules for the ordinary topology, into the associated sheaf of κ-modules for the (finer) hypercanonical topology. The "kernel" of this functor is exactly the category of torsion sheaves in the sense of [27], and this category is equivalent to the category $\underline{L}$ [cf. (7)] of torsion modules of R in the sense of [27].

Remark 2. - It is easily seen that $\underline{\mathrm{Mod}}_{X,h}(\kappa)$ is a discrete spectral category if and only if X comes from a complete atomic Boolean algebra, and in this case

$$\underline{\mathrm{Mod}}_{X,h}(\kappa) = \prod_{\substack{x \in X \\ x \text{ isolated}}} \mathrm{Mod}(\kappa(x))$$

Here $\kappa(x)$ denotes the stalk of κ over x.

Remark 3. - The hypercanonical topology on X is quite different from the ordinary topology, and in fact one can prove that $\underline{\mathrm{Mod}}_{X,h}(\kappa)$ can never be equivalent to the category of sheaves of modules over some sheaf of rings over some topological space, unless the given spectral category is discrete. (See [35], Théorème 2 for more general results.)

Using the Theorem 3, it is now easy to extend the Theorem 3 of [34] for locally idempotent categories to locally distributive categories and we can also deduce (cf. [35]):

THEOREM 4. - Let $\underline{C}$ be a locally distributive spectral category, whose modular dimension $\mu(\underline{C})$ satisfies $\mu(\underline{C}) \leq 2$ [33], [35]. Then $\underline{C}$ is discrete.

Remark 1. - It is quite probable, that even $\mu(\underline{C}) < \infty$ implies that $\underline{C}$ is discrete for locally distributive spectral categories $\underline{C}$. However, even to decide whether Theorem 4 is true or false for general spectral categories seems to be a difficult problem.

From Theorem 3 above, it is also easy to deduce the following result on the part R_1 of a right self-injective regular ring (cf. Corollary 3 of Theorem 1).

COROLLARY TO THEOREM 3. - Let X be a complete Boolean space, κ a sheaf of skewfields over X that is also a sheaf for the hypercanonical topology, and ν a sheaf of κ-modules that is also a sheaf for the hypercanonical topology. Then the endomorphism ring of ν is the same in $Mod_X(\kappa)$ as in $Mod_{X,h}(\kappa)$ and it is a locally strongly regular right self-injective ring, that is self-injective to the left too if and only if the open set $\{x \mid \dim_{\kappa(x)} \nu(x) < \infty\}$ is dense in X. In general, the left injective enveloping ring $_{\cdot}\overline{R}$ of R (cf. § 1) is isomorphic to the endomorphism ring of the sheaf of left κ-modules $\mathcal{H}om_{\kappa}(\nu,\kappa)$. Conversely, every locally strongly regular right self-injective ring can be obtained in the manner described in this corollary.

Remark 2. - We leave to the reader to formulate the preceeding corollary in the discrete case (compare Remark 2 following Theorem 3 above and [35]).

§ 4. - Example: The category of abelian sheaves on a topological space has a locally distributive spectral category.

We start with a general result about families of generators in a spectral category, thereby correcting an assertion in [12] p. 391.

PROPOSITION 4. - Let $\underline{C}$ be a Grothendieck category, $\{U_\alpha\}_{\alpha \in I}$ a family of generators of $\underline{C}$, and $\underline{C} \xrightarrow{P} \mathrm{Spec}(\underline{C})$ the natural functor from $\underline{C}$ to its spectral category. Then

$$\{P(U_\alpha/V) \mid V \text{ subobject of } U_\alpha,\ \alpha \in I\}$$

is a family of generators for $\mathrm{Spec}(\underline{C})$.

Remark. - If $\underline{C} = \mathrm{Mod}(\mathbb{Z})$, then $P(\mathbb{Z})$ is not a generator for $\mathrm{Spec}(\underline{C})$, so we can not expect a simpler result than Proposition 4.

PROOF OF PROPOSITION 4. - Let C be an arbitrary object of $\underline{C}$. Since the $\{U_\alpha\}$ form a family of generators, we see that we can construct by a transfinite argument an essential monomorphism

$$\coprod_{\gamma \in K} U_{\alpha_\gamma}/C_\gamma \longrightarrow C$$

But P transforms essential monomorphisms into isomorphisms and commutes with $\coprod$ (cf. § 1). Thus

$$\coprod_{\gamma \in K} P(U_{\alpha_\gamma}/C_\gamma) \xrightarrow{\simeq} P(C)$$

and so the Proposition 4 is proved.

THEOREM 5. - Let X be a topological space, and $\underset{\sim}{Z}$ the constant sheaf of the integers on X. Then $\mathrm{Spec}(\underline{\mathrm{Mod}}_X(\underset{\sim}{Z}))$ is a locally distributive spectral category. The same assertion is also true if we replace the constant sheaf $\underset{\sim}{Z}$ by a sheaf of skewfields over X.

PROOF: For every open set U in X, we denote by $\underset{\sim}{Z}_U$ the sheaf that is equal to $\underset{\sim}{Z}$ over U and zero elsewhere [14], [15]. It is known that the $\underset{\sim}{Z}_U$ form a family of generators of $\underline{\mathrm{Mod}}_X(\underset{\sim}{Z})$, and by Proposition 4 everything is proved, if we can prove that for every abelian subsheaf C of $\underset{\sim}{Z}_U$ we have that $P(\underset{\sim}{Z}_U/C)$ is distributive. We will prove an even stronger result [cf. § 1 and Proposition 2] namely:

(* *) If F_1 and F_2 are two abelian subsheaves of $\underset{\sim}{Z}_U/C$, with $F_1 \cap F_2 = 0$, then $\mathrm{Hom}_{\underline{C}}(F_1,F_2) = 0$.

But this is easy, for if $F_1 \xrightarrow{f} F_2$ were a non-zero morphism, then there would exist a point $x \in X$ where $F_1(x) \xrightarrow{f(x)} F_2(x)$ (the stalk morphism) would be a non-zero map of abelian groups. But

$F_1(x) \amalg F_2(x) \hookrightarrow \underset{\sim}{Z}_U(x)/C(x)$ and $\underset{\sim}{Z}_U(x)/C(x)$ is a cyclic abelian group, and so $F_1(x) \amalg F_2(x)$ is so too. But if E is a cyclic abelian group and $E = E_1 \amalg E_2$, then $\mathrm{Hom}_{\underset{\sim}{Z}}(E_1,E_2) = 0$ as is easily seen. Thus (* *) is verified and so $\mathrm{Spec}(\underline{\mathrm{Mod}}_X(\underset{\sim}{Z}))$ is locally distributive. The proof for a sheaf of skewfields is even simpler.

Remark 1. - If $\mathcal{A}$ is a sheaf of rings on a topological space X, then it is not true in general that $\mathrm{Spec}(\underline{\mathrm{Mod}}_X(\mathcal{A}))$ is locally distributive, even if X is a point, for there are rings A such that Spec(Mod(A)) is not locally distributive (cf. § 6).

Remark 2. - Comparing Theorem 5 with Corollary 2 of Theorem 1 we see in particular that to every topological space X and to every skewfield K, there is associated a uniquely determined strongly regular self-injective ring $R = R(K,X)$ that describes completely how the injective K-sheaves on X behave up to essential monomorphisms. The structure of this ring R is more complicated than we thought earlier. Consider the special case $K = \mathbb{Z}_2$, which has essentially all the features of the general case. Then R is a complete Boolean ring. The map

$$\underline{\mathrm{Mod}}_X(K) \xrightarrow{P} \mathrm{Spec}(\underline{\mathrm{Mod}}_X(K)) \quad (K = \mathbb{Z}_2)$$

transforms the object K_X into a distributive object (here even an idempotent object) whose endomorphism ring is easily seen to be isomorphic to the Boolean ring of regular open sets of X [14], [16]. **This** follows in fact from the theory of § 1 and Glivenko's theorem ([4] p. 216). But $P(K_X)$ is not the maximal distributive object of $\mathrm{Spec}(\underline{\mathrm{Mod}}_X(K))$. In fact, if S is a locally closed subset of X, then $P(K_S)$ is of course distributive and further

$$\mathrm{Hom}_{\mathrm{Spec}(\underline{\mathrm{Mod}}_X(K))}(P(K_X), P(K_S)) = 0$$

if and only if the complement of S is dense in X, so in this case $P(K_X) \amalg P(K_S)$ is also distributive by the Corollary 1 of Proposition 2. In particular, if X is a Hausdorff space (more generally, if every point of X is closed) then $\mathrm{Hom}(P(K_X), P(K_{\{x\}})) = 0$ for all non-isolated points x of X, and then it follows that

$$(* * *) \qquad P(K_X) \amalg \coprod_{\substack{x \in X \\ x \text{ non-} \\ \text{isolated}}} P(K_{\{x\}})$$

is a distributive object in $\mathrm{Spec}(\underline{\mathrm{Mod}}_X(K))$. But even this is not the maximal distributive object, except in special cases. If $X = [0,1] \subset \mathbb{R}$, then we can also add $P(K_F)$ to (* * *), where F is the ternary Cantor set, and still get a distributive object ...

Remark 3. - It seems to be interesting to study $\mathrm{Spec}(\underline{\mathrm{Mod}}_X(\mathcal{A}))$ where

$\mathcal{A}$ is a sheaf of rings on X, at least in the following cases:

1) $\mathcal{A}$ is the sheaf of germs of continuous functions (real, complex ...) on X.

2) $\mathcal{A}$ is the sheaf of germs of C^∞- functions on a differentiable manifold (or space).

3) X is a (not necessarily locally noetherian) scheme and $\mathcal{A} = 0_X$ = the structure sheaf on X.

4) The analytic variant of 3).

In case 1) one can easily see that the endomorphism ring of $P(\mathcal{A})$ is the maximal ring of quotients [10] of the ring of (real, complex ...) continuous functions on X. This ring has been studied in [10]. But from the theory developed here, one sees immediately that this is a strongly regular self-injective ring, and so one can apply § 3. We hope to return to this later. In case 3) the results of Hartshorne [17] give that the spectral category is discrete when X is noetherian, but this is not so in general, as follows from § 3.

§ 5. - On the endomorphism rings of injective objects in Grothendieck categories.

Let $\underline{C}$ be a Grothendieck category and I an injective object of $\underline{C}$. Then for every essential monomorphism $V \hookrightarrow I$, the natural restriction map $\mathrm{Hom}_{\underline{C}}(I,I) \to \mathrm{Hom}_{\underline{C}}(V,I)$ is an epimorphism (this is even true if I is only quasi-injective, and in fact can be taken as a definition of this concept). Passing to the direct limit we find that P defines a natural epimorphism

$$(9) \qquad \mathrm{Hom}_{\underline{C}}(I,I) \to \mathrm{Hom}_{\mathrm{Spec}(\underline{C})}(P(I),P(I))$$

It is clear that (9) is a ring map, and so the kernel of (9) is a two-sided ideal of $\mathrm{Hom}_{\underline{C}}(I,I)$. It is well-known that this kernel is the Jacobson radical of $\mathrm{Hom}_{\underline{C}}(I,I)$ (for a proof see e.g. [29] which works here too). Summing up we obtain:

THEOREM 6. - Let I be an injective (or even a quasi-injective) object in a Grothendieck category. Then the ring $\mathrm{Hom}_{\underline{C}}(I,I)/\mathrm{rad}_J(\mathrm{Hom}_{\underline{C}}(I,I))$ (here $\mathrm{rad}_J(\ )$ denotes the Jacobson radical) is naturally isomorphic to the endomorphism ring of the image of I in the spectral category of $\underline{C}$.

COROLLARY 1. - The quotient ring of Theorem 6 is von Neumann regular and right self-injective.

PROOF: By the theory of Gabriel-Oberst (cf. § 1 and [12]), the endomorphism ring of P(I) is regular and right self-injective.

Remark 1. - That the ring in Corollary 1 is regular has been known for some time [42] but the fact that it is right self-injective has only been proved recently in a quite different way in the module case by Osofsky [24] and Renault [30]. It should be remarked that using [28] one can reduce the general case to the module case.

COROLLARY 2. - If P(I) is a semi-simple object, then the ring in Theorem 6 is isomorphic to a product of endomorphism rings of vector spaces over skewfields.

Remark 2. - This has been proved in another way in [7] for modules. In particular, if $\underline{C}$ is locally coirreducible [22], then we can apply Corollary 2 to all injectives of $\underline{C}$.

COROLLARY 3. - Let X be a topological space, and I an injective abelian sheaf on X. Then $\mathrm{Hom}_{\underline{C}}(I,I)/\mathrm{rad}_J(\mathrm{Hom}_{\underline{C}}(I,I))$ is a locally strongly regular right self-injective ring.

Remark 3. - Much more precise results about $\mathrm{Hom}_{\underline{C}}(I,I)$ can be obtained for arbitrary $\underline{C}$ if we suppose that P commutes with filtered $\varinjlim$ and/or with arbitrary $\varprojlim$. In the case where also $\mathrm{Spec}(\underline{C})$ is discrete, we can just invoke [11] (cf. A-C of § 1), but there are also continuous analogues of Gabriel's results (continuous Krull dimension etc....).

§ 6. Remarks on the structure of general spectral categories.

We remarked in § 4 that even the spectral category of a module category is not necessarily locally distributive. The simplest example of this seems to be the following (cf. [36] and the literature cited there):

Let K be a skewfield and let R be the regular ring $\varinjlim_n M_{2^n}(K)$, where $M_{2^n}(K)$ denotes the ring of $2^n \times 2^n$ matrices with coefficients in K, and where the maps $M_{2^n}(K) \to M_{2^{n+1}}(K)$ are defined by $M \longmapsto \begin{pmatrix} M & 0 \\ 0 & M \end{pmatrix}$. Let $\bar{R}$ denote the injective envelope of R on any side (cf. § 1). To take the injective envelope is here the same as to complete R for the rank topology of [23] and the resulting regular ring $\bar{R}$ is both left <u>and</u> right self-injective (this phenomenom is closely related to the appearance of continuous geometries....). Let $\underline{C}_{\bar{R}}$ be the spectral category associated to $\bar{R}$. This category is <u>not</u> locally distributive, and neither is Spec(Mod(R)) that has $\underline{C}_{\bar{R}}$ as a direct factor. However $\underline{C}_{\bar{R}}$ is indecomposable [36], and furthermore it satisfies the axiom AB 5^*_{loc} of [36], i.e. it has a family of generators (here even <u>one</u> generator) $\{U_\alpha\}$ such that every U_α is an AB 5^*-object. Recall that an AB 5^*-object [36] is an object C such that for every $D \subset C$ and for every filtered decreasing family of subobjects of C, $\{C_\alpha\}$, the natural map

$$(\cap C_\alpha) + D \longrightarrow \cap (C_\alpha + D)$$

is an isomorphism. It is easy to see that an anti-distributive spectral category $\underline{C}$ decomposes in an essentially unique way into a direct product

$$\underline{C}_{adistr} \xrightarrow{\sim} \underline{C}_{II} \times \underline{C}_{III}$$

where $\underline{C}_{II}$ satisfies the axiom AB 5^*_{loc} and where $\underline{C}_{III}$ has no non-trivial AB 5^*-objects.

Thus we have three main types of spectral categories:

Type I = <u>the locally distributive spectral categories</u>,

Type II = <u>the anti-distributive spectral categories that satisfy</u> AB 5^*_{loc},

Type III = <u>the anti-distributive spectral categories that have no non-trivial</u> AB 5^*_{loc}<u>-object</u>, and we have just shown that type II can occur. This is also the case for type III [36]. As for the structure theory of type II and III, we refer the reader to our note [36] (Problème 1 of this note must be decomposed into two problems concerning type II and type III in order not to have an evident negative solution).

It seems now quite probable that even a more explicit structure theory for type II and III spectral categories, analogous to that for type I can be developed, the difference from the case I seems to be that the sheaf of skewfields over a complete Boolean space should be replaced by a sheaf of rings, whose stalks are coordinatizing rings of irreducible non-discrete continuous geometries, resp. upper-continuous geometries. Furthermore, it seems that Problème 1 (corrected as indicated above) could be reduced to the irreducible case.

Anyhow, the results of [36] make an extensive study of irreducible (upper)-continuous geometries highly desirable, and we hope to return to this problem soon.

Nice examples of (upper)-continuous geometries will probably be given by the study of Spec(Mod(A)), when A is a Banach algebra (we forget the topology), and there seems to be a definite connection between our types I-III for spectral categories, and the types I-III of continuous linear representations of topological groups (see e.g. Mackey [19]).

REFERENCES

[1] R. ARENS and I. KAPLANSKY, Topological representations of algebras, Trans. Amer. Math. Soc., 63, 1948, p. 457 - 481.

[2] M. ARTIN, Grothendieck Topologies, Harvard University, 1962.

[3] M. AUSLANDER, On regular group rings, Proc. Amer. Math. Soc., 8, 1957, p. 658 - 664.

[4] G. BIRKHOFF, Lattice Theory, Amer. Math. Soc. Coll. Publ. n° 25, 3rd edition, 1967.

[5] N. BOURBAKI, Algèbre commutative, Chap. I-II, Hermann, Paris, 1961.

(continued)

[6] B. BRAINERD and J. LAMBEK, On the ring of quotients of a Boolean ring, Can. Math. Bull., 2, 1959, p. 25 - 29.

[7] A. CAILLEAU, Anneau associé à un module injectif riche en coirréductibles, Comptes rendus, série A, 264, 1967, p. 1040 - 1042.

[8] J. DAUNS and K.H. HOFMANN, The representation of biregular rings by sheaves, Math. Z., 91, 1966, p. 103 - 123.

[9] S. EILENBERG - J.C. MOORE, Foundations of Relative Homological Algebra, Memoirs of the Amer. Math. Soc. n° 55, 1965.

[10] N.J. FINE, L. GILLMAN and J. LAMBEK, Rings of Quotients of Rings of Functions, McGill University Press, Montreal, 1966.

[11] P. GABRIEL, Des catégories abéliennes, Bull. Soc. Math. France, 90, 1962, p. 323 - 448.

[12] P. GABRIEL and U. OBERST, Spektralkategorien und reguläre Ringe im von Neumannschen Sinn, Math. Z., 92, 1966, p. 389 - 395.

[13] J. GIRAUD, Analysis situs, Séminaire Bourbaki, 15, 1962-1963, Exposé 256.

[14] R. GODEMENT, Théorie des faisceaux, Hermann, Paris, 1958.

[15] A. GROTHENDIECK, Sur quelques points d'algèbre homologique, Tohoku Math. J., 9, 1957, p. 119 -221.

[16] P.R. HALMOS, Lectures on Boolean Algebras, van Nostrand, Princeton, 1963.

[17] R. HARTSHORNE, Residues and Duality, Springer, Berlin, 1966.

[18] C.U. JENSEN, Arithmetical rings, Acta Math. Acad. Scient. Hung., 17, 1966, p. 115 - 123.

[19] G.W. MACKEY, The Theory of Group Representations, University of Chicago, 1955.

[20] S. MACLANE, Homology, Springer, Berlin, 1963.

[21] D. MUMFORD, Picard groups of moduli problems, p. 33 - 81 in: Arithmetical Algebraic Geometry, Harper and Row, New York, 1965.

[22] C. NĂSTĂSESCU - N. POPESCU, Quelques observations sur les topos abéliens, Rev. Roum. Math. Pures et Appl., 12, 1967, p. 553 - 563.

[23] J. von NEUMANN, Continuous Geometry, Princeton University Press, Princeton, 1960.

(continued)

[24] B. OSOFSKY, Endomorphism rings of quasi-injective modules, Notices Amer. Math. Soc., 14, 1967, p. 419.

[25] R.S. PIERCE, Distributivity in Boolean algebras, Pacific J. Math., 7, 1957, p. 983 - 992.

[26] R.S. PIERCE, Distributivity and the normal completion of Boolean algebras, Pacific J. Math., 8, 1958, p. 133 - 140.

[27] R.S. PIERCE, Modules over Commutative Regular Rings, Memoirs of the Amer. Math. Soc. n° 70, 1967.

[28] N. POPESCU and P. GABRIEL, Caractérisation des catégories abéliennes avec générateurs et limites inductives exactes, Comptes rendus, 258, 1964, p. 4188 - 4190.

[29] G. RENAULT, Anneaux self-injectifs, Séminaire Dubreil-Pisot, 19, 1965 - 1966, Exposé 11.

[30] G. RENAULT, Anneau associé à un module injectif, Comptes rendus, 264, série A, 1967, p. 1163 - 1164.

[31] G. RENAULT, Etude des sous-modules compléments dans un module, Mémoires de la Soc. Math. de France, n° 9, 1967.

[32] J.-E. ROOS, Introduction à l'étude de (resp. Sur, resp. Complément à l'étude de) la distributivité des foncteurs $\varprojlim$ par rapport aux $\varinjlim$ dans les catégories des faisceaux (topos), Comptes rendus, 259, 1964, p. 969 - 972, p. 1605 - 1608 et p. 1801 - 1804.

[33] J.-E. ROOS, Caractérisation des catégories qui sont quotients de catégories de modules par des sous-categories bilocalisantes, Comptes rendus, 261, 1965, p. 4954 - 4957.

[34] J.-E. ROOS, Sur la condition AB 6 et ses variantes dans les catégories abéliennes, Comptes rendus, série A, 264, 1967, p. 991 - 994.

[35] J.-E. ROOS, Sur les catégories spectrales localement distributives, Comptes rendus, série A, 265, 1967, p. 14 - 17.

[36] J.-E. ROOS, Sur la structure des catégories spectrales et les coordonnées de von Neumann des treillis modulaires et complémentés, Comptes rendus, série A, 265, 1967, p. 42 - 45.

[37] Séminaire M. ARTIN - A. GROTHENDIECK, I.H.E.S., 1963 - 1964, fasc. 1, Exposés 1 - 6 (by J.-L. VERDIER).

[38] R. SIKORSKI, Boolean algebras, 2nd edition, Springer, Berlin, 1964.

[39] L.A. SKORNJAKOV, Elizarovskoe kol'co častnyh i princip lokalisacii, Matematičeskie Zametki, 1, 1967, p. 263 - 268.

(continued)

[40] E.C. SMITH Jr and A. TARSKI, Higher degrees of distributivity and completeness in Boolean algebras, Trans. Amer. Math. Soc., 84, 1957, p. 230 - 257.

[41] B. STENSTRÖM, High submodules and purity, Arkiv för matematik, 7, 1967, p. 173 - 176.

[42] Y. UTUMI, On continuous rings and self-injective rings, Trans. Amer. Math. Soc., 118, 1965, p. 158 - 173.

ecture Notes in Mathematics

r erschienen/Already published

J. Wermer, Seminar über Funktionen-Algebren.
Seiten. 1964. DM 3,80

A. Borel, Cohomologie des espaces localement
cts d'après J. Leray.
pages. 1964. DM 9,–

J. F. Adams, Stable Homotopy Theory.
vised edition. IV, 78 pages. 1966. DM 7,80

M. Arkowitz and C. R. Curjel, Groups of Homotopy
s. 2nd. revised edition. IV, 36 pages. 1967.
30

J.-P. Serre, Cohomologie Galoisienne.
me édition. VIII, 214 pages. 1965. DM 18,–

H. Hermes, Eine Termlogik mit Auswahloperator.
Seiten. 1965. DM 5,80

Ph. Tondeur, Introduction to Lie Groups
ansformation Groups.
5 pages. 1965. DM 13,50

G. Fichera, Linear Elliptic Differential
s and Eigenvalue Problems.
pages. 1965. DM 13.50

P. L. Ivănescu, Pseudo-Boolean Programming and
tions. IV, 50 pages. 1965. DM 4,80

H. Lüneburg, Die Suzukigruppen und ihre
rien. VI, 111 Seiten. 1965. DM 8,–

J.-P. Serre, Algèbre Locale. Multiplicités.
par P. Gabriel. Seconde édition.
pages. 1965. DM 12,–

A. Dold, Halbexakte Homotopiefunktoren.
Seiten. 1966. DM 12,–

E. Thomas, Seminar on Fiber Spaces.
pages. 1966. DM 4,80

H. Werner, Vorlesung über Approximations-
IV, 184 Seiten und 12 Seiten Anhang. 1966. DM 14,–

F. Oort, Commutative Group Schemes.
pages. 1966. DM 9,80

Vol. 16: J. Pfanzagl and W. Pierlo, Compact Systems of Sets. IV, 48 pages. 1966. DM 5,80

Vol. 17: C. Müller, Spherical Harmonics. IV, 46 pages. 1966. DM 5,–

Vol. 18: H.-B. Brinkmann und D. Puppe, Kategorien und Funktoren. XII, 107 Seiten. 1966. DM 8,–

Vol. 19: G. Stolzenberg, Volumes, Limits and Extensions of Analytic Varieties. IV, 45 pages. 1966. DM 5,40

Vol. 20: R. Hartshorne, Residues and Duality. VIII, 423 pages. 1966. DM 20,–

Vol. 21: Seminar on Complex Multiplication. By A. Borel, S. Chowla, C. S. Herz, K. Iwasawa, J.-P. Serre. IV, 102 pages. 1966. DM 8,–

Vol. 22: H. Bauer, Harmonische Räume und ihre Potentialtheorie. IV, 175 Seiten. 1966. DM 14,–

Vol. 23: P. L. Ivănescu and S. Rudeanu, Pseudo-Boolean Methods for Bivalent Programming. 120 pages. 1966. DM 10,–

Vol. 24: J. Lambek, Completions of Categories. IV, 69 pages. 1966. DM 6,80

Vol. 25: R. Narasimhan, Introduction to the Theory of Analytic Spaces. IV, 143 pages. 1966. DM 10,–

Vol. 26: P.-A. Meyer, Processus de Markov. IV, 190 pages. 1967. DM 15,–

Vol. 27: H. P. Künzi und S.T. Tan, Lineare Optimierung großer Systeme. VI, 121 Seiten. 1966. DM 12,–

Vol. 28: P. E. Conner and E. E. Floyd, The Relation of Cobordism to K-Theories. VIII, 112 pages. 1966. DM 9.80

Vol. 29: K. Chandrasekharan, Einführung in die Analytische Zahlentheorie. VI, 199 Seiten 1966. DM 16.80

Vol. 30: A. Frölicher and W. Bucher, Calculus in Vector Spaces without Norm. X, 146 pages. 1966 DM 12,–

Bitte wenden/Continued

Vol. 31: Symposium on Probability Methods in Analysis. Chairman: D.A.Kappos. IV, 329 pages. 1967. DM 20,–

Vol. 32: M. André, Méthode Simpliciale en Algèbre Homologique et Algèbre Commutative. IV, 122 pages. 1967. DM 12,–

Vol. 33: G. I. Targonski, Seminar on Functional Operators and Equations. IV, 112 pages. 1967. DM 10,–

Vol. 34: G. E. Bredon, Equivariant Cohomology Theories. VI, 64 pages. 1967. DM 6,80

Vol. 35: N. P. Bhatia and G. P. Szegö, Dynamical Systems: Stability Theory and Applications. VI, 416 pages. 1967. DM 24,–

Vol. 36: A. Borel, Topics in the Homology Theory of Fibre Bundles. VI, 95 pages. 1967. DM 9,–

Vol. 37: R. B. Jensen, Modelle der Mengenlehre. X, 176 Seiten. 1967. DM 14,–

Vol. 38: R. Berger, R, Kiehl, E. Kunz und H.-J. Nastold, Differentialrechnung in der analytischen Geometrie. IV, 134 Seiten. 1967. DM 12,–

Vol. 39: Séminaire de Probabilités I. IV, 187 pages. 1967. DM 14,–

Vol. 40: J. Tits, Tabellen zu den einfachen Lie Gruppe und ihren Darstellungen. VI, 53 Seiten. 1967. DM 6,80

Vol. 41: R. Hartshorne, Local Cohomology. VI, 106 pages. 1967. DM 10,–

Vol. 42: J. F. Berglund and K. H. Hofmann, Compact Semitopological Semigroups and Weakly Almost Peri Functions. VI, 160 pages. 1967. DM 12,–

Vol. 43: D. G. Quillen, Homotopical Algebra. VI, 157 pages. 1967. DM 14,–

Vol. 44: K. Urbanik, Lectures on Prediction Theory. II, 50 pages. 1967. DM 5,80

Vol. 45: A. Wilansky, Topics in Functional Analysis. VI, 102 pages. 1967. DM 9,60

Vol. 46: P. E. Conner, Seminar on Periodic Maps. IV, 116 pages. 1967. DM 10,60